Virus Bioinformatics

Chapman & Hall/CRC Computational Biology Series

About the Series:

This series aims to capture new developments in computational biology, as well as high-quality work summarizing or contributing to more established topics. Publishing a broad range of reference works, textbooks, and handbooks, the series is designed to appeal to students, researchers, and professionals in all areas of computational biology, including genomics, proteomics, and cancer computational biology, as well as interdisciplinary researchers involved in associated fields, such as bioinformatics and systems biology.

PUBLISHED TITLES

Clustering in Bioinformatics and Drug Discovery
John David MacCuish and Norah E. MacCuish

Metabolomics: Practical Guide to Design and Analysis
Ron Wehrens and Reza Salek

An Introduction to Systems Biology: Design Principles of Biological Circuits
Second Edition
Uri Alon

Computational Biology: A Statistical Mechanics Perspective
Second Edition
Ralf Blossey

Stochastic Modelling for Systems Biology
Third Edition
Darren J. Wilkinson

Computational Genomics with R
Altuna Akalin, Bora Uyar, Vedran Franke, and Jonathan Ronen

An Introduction to Computational Systems Biology: Systems-level Modelling of Cellular Networks
Karthik Raman

Virus Bioinformatics
Dmitrij Frishman and Manja Marz

For more information about this series please visit:
https://www.routledge.com/Chapman--HallCRC-Computational-Biology-Series/book-series/CRCCBS

Virus Bioinformatics

Edited by
Dmitrij Frishman
Manja Marz

CRC Press
Taylor & Francis Group
Boca Raton London New York

CRC Press is an imprint of the
Taylor & Francis Group, an **informa** business

A CHAPMAN & HALL BOOK

First edition published 2022
by CRC Press
6000 Broken Sound Parkway NW, Suite 300, Boca Raton, FL 33487-2742

and by CRC Press
2 Park Square, Milton Park, Abingdon, Oxon, OX14 4RN

Library of Congress Cataloging-in-Publication Data
Names: Frishman, Dmitrij, editor. | Marz, Manuela, editor.
Title: Virus bioinformatics / edited by Dmitrij Frishman, Manuela Marz.
Description: First edition. | Boca Raton : CRC Press, 2021. |
Series: Chapman & Hall/CRC computational biology series |
Includes bibliographical references and index.
Identifiers: LCCN 2021008505 | ISBN 9780367558604 (hardback) |
ISBN 9780367564193 (paperback) | ISBN 9781003097679 (ebook)
Subjects: LCSH: Virology—Data processing. | Virology—Research. |
Bioinformatics.
Classification: LCC QR370.V57 2021 | DDC 579.0285—dc23
LC record available at https://lccn.loc.gov/2021008505

ISBN: 978-0-367-55860-4 (hbk)
ISBN: 978-0-367-56419-3 (pbk)
ISBN: 978-1-003-09767-9 (ebk)

Typeset in Minion Pro
by codeMantra

Contents

Preface

THIS BOOK WAS CONCEIVED LONG BEFORE THE START OF THE COVID-19 pandemic. While there are multiple excellent publications covering various aspects of bioinformatics—from sequence analysis and structural bioinformatics to regulation and genome evolution—we felt that a dedicated book on bioinformatics of viruses would be warranted given their remarkable and very special properties, which render many standard algorithms and methods inefficient and dictate the application of highly specialized tools. As this book progressed toward completion, the actual scale of the COVID-19 pandemic became apparent, giving further emphasis to the importance of the field. Indeed, at least half of the chapters in this book mention the SARS-COV-2 virus, and two chapters are dedicated to it.

We have described only a tiny fraction of the estimated total number of viruses (<5000, estimated to be <1% of the total), and we can rarely claim to fully understand even these. The community of virologists is itself relatively small. If we want to drive the field forward, we need to connect it with other disciplines. Perhaps the most promising of these is bioinformatics, but there are unique challenges within this field. There is no unified database for viruses; currently, we have only a few databases for specific viruses such as Influenza, HIV, and human pathogenic viruses. Further, while the genomes of viruses are relatively small, consisting of only a few genes, the number of sequences is huge (if you wanted to build an alignment of the Influenza A virus, for example, you would need to align several hundred thousand sequences). The high mutation rate also makes it difficult to explore the complete space of genomic sequences.

When we follow a replication cycle of viruses, we discover further bioinformatical challenges. Our virus might be outside the host—usually it appears in various haplotypes of a quasispecies—and it is still an unsolved problem to describe the whole quasispecies, its space of genomic variants, and fitness. We may consider a single particle, but we still do not know the genomic arrangement within the particle. Another problem arises when the virus enters a host: there are around ten different ways of entry described and most of these are not understood. Then, after a virus enters a host, several interactions happen, some at a DNA level, some at an RNA level. With general high-throughput transcriptomic approaches, we start to understand the reaction of the host. However, the specific processes for replication and the interaction with the host are poorly understood. The "assembly" of viruses (virologists use the word "assembly" in a different sense to bioinformaticians) is also a mystery in many cases.

Finally, we also lack an understanding of the history of viruses. Many virologists agree that viruses do not have a common ancestor, but rather several origins. We must therefore understand the coevolution of viruses and hosts. In addition, viruses have a high mutation rate and a fast turn-over rate, which makes them a good model for fast real-time evolution. In fact, by learning more about viruses, we could make completely new discoveries in both molecular biology and evolution.

This book aims to provide an overview of the state-of-the-art virus bioinformatics tools as well as current research within the field. The first three chapters of this book are devoted to different aspects of *virus genomics*. In Chapter 1, Thomas Rattei discusses comparative genomics of viruses. He describes the current state of the art in sequencing, assembling, and annotating viral genomes from cultures and environmental samples and then focuses on the difficult task of deriving viral orthologous families. In Chapter 2, Simon Roux and Mark Borodovsky give a comprehensive account of experimental and algorithmic challenges in phage metagenomics. Specialized tools for predicting genes in metagenome-assembled phage genomes are described that are capable of dealing with some of the intrinsic obstacles, such as small genome size, occurrence of overlapping genes, and programmed frameshifts, as well as unknown genetic code. In Chapter 3, Sebastian Krautwurst, Ronald Dijkman, Volker Thiel, Andi Krumbholz, and Manja Marz provide a thorough review of the emerging technology of direct RNA sequencing and its multifarious applications in virology, including tracking virus mutations during pandemics. They discuss the advantages of long reads for assembling viral genomes and for discovering long-range interactions, compensatory mutations, and RNA modifications.

In Chapter 4, Kim Philipp Jablonski and Niko Beerenwinkel summarize contemporary approaches to estimating the *genetic diversity* of viruses based on next-generation sequencing data and present a survey of de novo and reference-based algorithms for global haplotype reconstruction.

The next two chapters cover the area of *RNA informatics.* In Chapter 5, Michael Wolfinger, Roman Ochsenreiter, and Ivo Hofacker employ structure prediction and phylogenetic tools to investigate the role of RNA secondary structural elements in virus tropism, on the specific example of 3′ UTR regions of flaviviruses. In Chapter 6, Alexander Gultyaev, René Olsthoorn, Monique Spronken, and Mathilde Richard give an overview of RNA structures, including pseudoknots, in the influenza virus genomes identified both by computational methods and structure probing techniques, and discuss their evolutionary conservation and functional significance at various steps of the replication cycle.

Four chapters address *virus-host interactions.* In Chapter 7, Ziyang Gao, Senbao Lu, Oleksandr Narykov, Suhas Srinivasan, and Dmitry Korkin dissect the intricate network of intra-viral and virus-host interactions in the SARS-CoV-2 and its implications for finding drug targets. Using homology modeling and docking, they reconstruct protein-protein and protein-ligand complexes and discuss the evolutionary conservation of the binding sites and the possible consequences of their disruption by mutations. In Chapter 8, Hadi Karimzadeh, Daniel Habermann, Daniel Hoffmann, and Michael Roggendorf explore the bioinformatics methods that help to identify T-cell epitopes and immune escape mutations, using the hepatitis delta virus as a model. In Chapter 9, Caroline Friedel focuses on the methodological challenges of "dual RNA Seq," the parallel profiling of virus and host transcriptomes, and describes custom algorithms for read mapping, normalization, and differential expression as well as the issues associated with data curation and interpretation. In Chapter 10, Florian Mock and Manja Marz demonstrate how machine learning can be used to predict the most likely host organisms for viruses using their nucleotide sequences as a sole input, based on the example of the Influenza virus A.

Two chapters address specific aspects of *modeling and synthetic biology.* In Chapter 11, Manasa KP, Kamilya Altynbekova, and Alexander Kel explain how targets for drug repurposing can be identified by analyzing SARS-CoV-2 gene regulatory networks derived from transciptomics data. In Chapter 12, Henni Zommer and Tamir Tuller discuss the perspectives of tumor virotherapy and present the current state in mathematical modeling of oncolytic viruses and designing vaccines.

The final chapter of this book, by Edouard De Castro, Chantal Hulo, Patrick Masson, and Philippe Le Mercier, covers *digitalization and dissemination of knowledge* in virology. It gives an overview of the Web resources and databases relevant for virology researchers and zooms in on the ViralZone project, hosted at the Swiss Institute for Bioinformatics, which systematically collects molecular and epidemiologic information on viral protein families.

Viruses are the most numerous biological entities on the planet, infecting all types of living organisms—from bacteria to human beings. The constantly expanding repertoire of experimental approaches available to study viruses includes both low throughput techniques, such as imaging and 3D structure determination, and modern OMICS technologies, such as genome sequencing, ribosomal profiling, and transcriptome-wide RNA structure probing. Bioinformatics of viruses faces significant challenges due to their seemingly unlimited diversity, unusual lifestyle, great variety of replication strategies, compact genome organization, and rapid rate of evolution. At the same time, it also has the potential to deliver decisive clues for developing vaccines and medications against dangerous viral outbreaks. This book reviews state-of-the-art bioinformatics algorithms and recent advances in data analysis within virology. Leading experts in the field cover a variety of topics, ranging from virus genome analysis and RNA structure prediction to virus evolution and virus-host interactions. This book appeals to computational biologists wishing to venture into the rapidly advancing field of virus bioinformatics as well as to virologists interested in acquiring basic bioinformatics skills to support their wet lab work.

We hope this book will encourage many more to enter this crucial field. There is an endless amount left to explore.

Dmitrij Frishman and Manja Marz
Munich and Jena
01/26/2021

Editors

Dmitrij Frishman is Professor for Bioinformatics at Munich Technical University (TUM).

Manja Marz is Professor for High-Throughput Sequencing Analysis at Friedrich Schiller University Jena.

Contributors

Kamilya Altynbekova
geneXplain GmbH
Wolfenbüttel, Germany

Niko Beerenwinkel
Department of Biosystems Science
    and Engineering
ETH Zurich
Basel, Switzerland
and
SIB Swiss Institute of
    Bioinformatics
Basel, Switzerland

Mark Borodovsky
Wallace H. Coulter Department of
    Biomedical Engineering
Georgia Tech
Atlanta, Georgia
and
School of Computational Science
    and Engineering
Georgia Tech
Atlanta, Georgia

Edouard De Castro
Swiss Institute of Bioinformatics
Geneva, Switzerland

Ronald Dijkman
Institute of Virology and
    Immunology
University of Bern
Bern, Switzerland
and
European Virus Bioinformatics
    Center
Friedrich Schiller University
    Jena
Jena, Germany

Caroline C. Friedel
Ludwig-Maximilians-Universität
    München
Munich, Germany

Ziyang Gao
Bioinformatics and Computational
    Biology Program
Worcester Polytechnic Institute
Worcester, Massachusetts

Alexander P. Gultyaev
Group Imaging & Bioinformatics
Leiden Institute of Advanced
    Computer Science (LIACS)
Leiden University
Leiden, the Netherlands
and
Department of Viroscience
Erasmus Medical Center
Rotterdam, the Netherlands

Daniel Habermann
Department of Bioinformatics
University of Duisburg-Essen
Essen, Germany

Ivo L. Hofacker
Department of Theoretical
    Chemistry
University of Vienna
Vienna, Austria
and
Research Group BCB, Faculty of
    Computer Science
University of Vienna
Vienna, Austria

Daniel Hoffmann
Department of Bioinformatics
University of Duisburg-Essen
Essen, Germany

Chantal Hulo
Swiss Institute of Bioinformatics
Geneva, Switzerland

Kim Philipp Jablonski
Department of Biosystems Science
    and Engineering
ETH Zurich
Basel, Switzerland
and
SIB Swiss Institute of
    Bioinformatics
Basel, Switzerland

Hadi Karimzadeh
Division of Clinical Pharmacology
University Hospital, Ludwig
    Maximilian University
Munich, Germany

Alexander Kel
geneXplain GmbH
Wolfenbüttel, Germany
and
Laboratory of Pharmacogenomics
Institute of Chemical Biology and
    Fundamental Medicine SBRAS
Novosibirsk, Russia

Dmitry Korkin
Bioinformatics and Computational
    Biology Program
Worcester Polytechnic Institute
Worcester, Massachusetts
and
Computer Science Department
Worcester Polytechnic Institute
Worcester, Massachusetts
and
Data Science Program
Worcester Polytechnic Institute
Worcester, Massachusetts

Sebastian Krautwurst
RNA Bioinformatics and High-
  Throughput Analysis
Friedrich Schiller University Jena
Jena, Germany
and
European Virus Bioinformatics
  Center
Friedrich Schiller University Jena
Jena, Germany

Andi Krumbholz
Institute for Infection Medicine
University Medical Center
  Schleswig-Holstein
Christian-Albrecht University of
  Kiel
Kiel, Germany
and
Labor Dr. Krause & Kollegen MVZ
  GmbH
Kiel, Germany
and
European Virus Bioinformatics
  Center
Friedrich Schiller University Jena
Jena, Germany

Senbao Lu
Bioinformatics and Computational
  Biology Program
Worcester Polytechnic Institute
Worcester, Massachusetts

Manasa Kalya
geneXplain GmbH
Wolfenbüttel, Germany
and
Department of Medical
  Bioinformatics
University Medical Center
  Göttingen
Göttingen, Germany

Patrick Masson
Swiss Institute of Bioinformatics
Geneva, Switzerland

Florian Mock
RNA Bioinformatics and
  High-Throughput Analysis
Friedrich Schiller University Jena
Jena, Germany

Oleksandr Narykov
Computer Science Department
Worcester Polytechnic Institute
Worcester, Massachusetts

Roman Ochsenreiter
Department of Theoretical
  Chemistry
University of Vienna
Vienna, Austria

René C.L. Olsthoorn
Group Supramolecular &
  Biomaterials Chemistry
Leiden Institute of Chemistry
Leiden University
Leiden, the Netherlands

Philippe Le Mercier
SIB Swiss Institute of
  Bioinformatics
University of Geneva Medical
  School
Genève, Switzerland

Thomas Rattei
CUBE - Division of Computational
  Systems Biology
University of Vienna
Vienna, Austria

Mathilde Richard
Department of Viroscience
Erasmus Medical Center
Rotterdam, the Netherlands

Michael Roggendorf
Institute of Virology
School of Medicine
Technical University of Munich
Munich, Germany

Simon Roux
DOE Joint Genome Institute
Lawrence Berkeley National
  Laboratory
Berkeley, California

Monique I. Spronken
Department of Viroscience
Erasmus Medical Center
Rotterdam, the Netherlands

Suhas Srinivasan
Data Science Program
Worcester Polytechnic Institute
Worcester, Massachusetts

Volker Thiel
Institute of Virology and
  Immunology
University of Bern
Bern, Switzerland
and
European Virus Bioinformatics
  Center
Friedrich Schiller University Jena
Jena, Germany

Tamir Tuller
Department of Biomedical
  Engineering
Tel-Aviv University
Tel Aviv, Israel

Michael T. Wolfinger
Department of Theoretical
  Chemistry
University of Vienna
Vienna, Austria
and
Research Group BCB, Faculty of
  Computer Science
University of Vienna
Vienna, Austria

Henni Zommer
Department of Biomedical
  Engineering
Tel-Aviv University
Tel Aviv, Israel

CHAPTER 1

Comparative Genomics of Viruses

Thomas Rattei

University of Vienna

CONTENTS

1.1 GENOMICS OF VIRUSES

1.1.1 Genome Types, Sizes, and Nomenclature

All viruses carry genetic information, which is encoded in genomic sequences. As a consequence of their host-based replication cycles, viruses usually have small or even very small genomes. However, the largest virus genome sequence can comprise more than a million nucleotides, which already is a typical size for small prokaryotic genomes (Figure 1.1a). Most eukaryotic genomes are based on single-stranded RNA (ssRNA), whereas most viruses of Bacteria and Archaea consist of double-stranded DNA

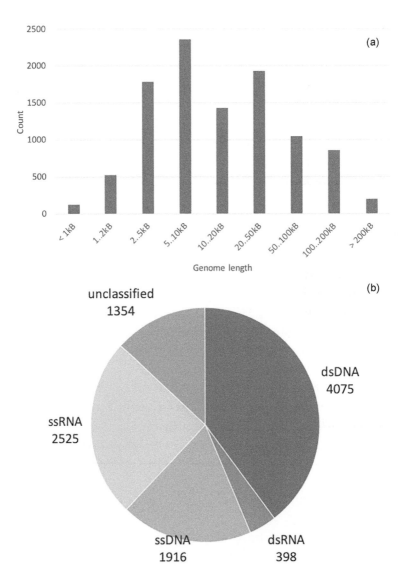

FIGURE 1.1 (a) Size distribution of virus genomes in NCBI RefSeq version 201 (O'Leary et al. 2016). (b) Genome types single-stranded RNA (ssRNA), double-stranded RNA (dsRNA), single-stranded DNA (ssDNA), and double-stranded DNA (dsDNA) in NCBI RefSeq version 202 (O'Leary et al. 2016).

(dsDNA) (Figure 1.1b). Despite their molecular structures, virus genome sequences are archived, exchanged, and computationally analyzed in the same one-letter IUPAC single-stranded nucleotide encoding as any cellular genome sequences. In the case of double-stranded genomes, canonical base pairing is assumed and one of the strands is selected for the genome sequence string. As both strands of a double-stranded genome are equivalent in terms of encoded information, the strand for the genome sequence of a newly sequenced virus genome is usually selected according to existing genome records in public sequence databases.

1.1.2 Genome Sequences from Cultures

Virus genomes can be sequenced from genetic material that is extracted from cultures of viruses in their host cells. These can be natural host cells as well as cell lines that are suitable for virus replication in the lab. The genomic material is extracted from the sample using standardized protocols. Ready-to-use kits for this purpose are commercially available for many viruses. However, their use for novel viruses often requires adaptation and validation. Virus genomic material is separated from host genomes and host transcripts by their different chemical makeup (e.g., single-stranded RNA vs. double-stranded DNA) and their different size and molecular weight, respectively.

The sequencing of virus genomes from cultures of many host cells rarely targets one uniform, static genome variant. Instead, a mixture of heterogeneous genome sequences is expected as a result of in-host evolution. This phenomenon is mostly remarkable in single-stranded RNA genomes, according to the limited error control during virus replication. Although the concept of "Quasispecies" initially described the effect of in-host evolution on fitness landscapes (Swetina and Schuster 1982), it is well-supported by recently collected genomic evidence from many viruses (Schuster 2016). Genome sequencing projects need to consider this phenomenon by their experimental design and their selection procedures for the genomic material. Specific genome assembly approaches allow the reconstruction of virus quasispecies genomes from deep short-read sequencing (Topfer et al. 2014). Recently developed sequencing techniques allow for long-read sequencing of complete virus genomes in single reads and provide a direct approach to the genome sequence diversity within a virus quasispecies (Yamashita et al. 2020).

Not always viral genomes are sequences on purpose. They can be sequenced along with host genomes and transcriptomes and are then usually removed and discarded. Furthermore, the presence of viruses in lab cell cultures and reagents is due to the abundance and diversity of viruses (Thannesberger et al. 2017).

1.1.3 Genomes from Environmental Samples

Virus genome sequences can also be obtained without cultivation, which is referred to as "metagenomics." There are many reasons for using this approach, such as the survey for unknown viruses, the assessment of natural in-host evolution, the attempt to quantify natural abundances, or simply the lack of a suitable cultivation method. Metagenomic sequencing is performed on the material extracted from an environmental sample, which includes isolates from single multicellular individuals. It can be applied to RNA and DNA viruses and results in a mixture of virus and cellular reads, depending on the extraction and separation protocols, sequencing technique, and sequencing depth (Greninger 2018, Schulz et al. 2020).

The computational analysis of viral metagenomes from short reads is usually performed in specific workflows. These first assemble the reads into contigs or scaffolds using assembly software that is aware of different abundance of reads from different species. Assemblies from different assemblers or from different samples can be merged into one single metagenome assembly (Olm et al. 2017). Scaffolds are grouped into metagenomic bins by their relative sequence read depth in different samples or different genome extractions and by the similarity of their oligonucleotide frequency profiles. Compared to the binning of cellular metagenomes, no universally conserved, single-copy marker genes can be used for the binning of virus metagenomic assemblies. Consequently, also no general approach for the assessment of completeness, heterogeneity and contamination of virus metagenomic bins could be developed so far. Minimum Information about any (x) Sequence (MIxS) standard has recently been developed for reporting sequences of uncultivated virus genomes (Roux et al. 2019).

1.1.4 Proviruses

A special group of viruses is proviruses, which are integrated into their host's genomes. Proviruses can be essential for the replication of viruses or can comprise latent forms of viruses. Both cases are relevant for virus genomics. Endogenous retroviruses make up significant portions of eukaryotic genomes, and prophages are frequently found in bacterial genomes. Proviruses are annotated according to their sequence characteristics in their host genomes, which can be combined with the prediction of whether the proviruses are still functional or degenerate. In genome assembly, the classification of viral contigs as provirus or viral contamination is challenging and requires the resolution of genomic repeats. Specific nomenclature for annotated retroviruses has been recently suggested (Gifford et al. 2018). Proviruses in microbial sequences can be automatically annotated based on their insertion sequence characteristics and their typical genome contents and gene order (Roux et al. 2015).

1.1.5 Annotation of Virus Genomes

The diversity of viral species, their life cycles, their genome structures, and their cultivability remain massive challenges for the development of universal software solutions for the annotation of virus genomes. Therefore, the main principles of automatic annotation of virus genomes are the detection of coding sequences by their oligonucleotide (such as codon) frequencies as well as the homology-based transfer of features and functional classifications from annotated genomes to newly sequenced genomes (Shean et al. 2019).

1.1.6 Database Resources for Virus Genome Sequences

The International Nucleotide Sequence Database Collaboration (INSDC) organizes the database resources that store newly sequenced and published genome sequences. It is a joint initiative of DDBJ, EMBL-EBI, and NCBI. INSDC has defined database and record structures for annotated genomes as well as partial genomic sequences, raw assemblies, and unassembled sequence reads. It includes formats for the attachment of functional annotation as well as contextual information relating to samples and experimental configurations (Cochrane et al. 2016).

Further databases with particular importance for virus genomes are NCBI RefSeq (O'Leary et al. 2016) and Uniprot/Swissprot (UniProt 2019). Whereas RefSeq is specialized in the selection and representation of complete genome sequences, SwissProt makes massive efforts in the manual curation of the annotations of viral gene products. Along with these efforts, ViralZone has been developed as a user-friendly knowledge base about virus genomics, including their virion structure, replication cycle, and host-virus interactions (Masson et al. 2013).

1.2 COMPARISON OF VIRUS GENOME SEQUENCES

The most direct approach to comparative genomics of viruses is the direct comparison of genome sequences to each other. This can be performed for whole genomes, genome fragments, short subsequences, and single nucleotides (Figure 1.2).

For comparisons of whole genomes, alignment-based methods and alignment-free methods exist. Alignment-based tools typically calculate regions of local similarity and subsequently extend these into whole genome alignments and visualizations (Darling et al. 2004). Alignments of partial or complete genomes can be calculated using generic methods, which utilize short common subsequences (Marcais et al. 2018). In order to account for the dynamics of genome evolution, which quickly leads to genome sequence divergence, additional constraints can be exploited to improve the accuracy of genome alignments, such as the sequence of codons and their encoded amino acids (Libin et al. 2019). For the comparison of many genomes, the similarities can be expressed numerically, e.g., as average nucleotide identity values (Varghese et al. 2015).

Alignment-free methods for genome alignment can focus on the visualization of genome similarities, which is particularly helpful to intuitively understand phenomena of genome evolution, such as transversions, inversions, and duplications (Krumsiek, Arnold, and Rattei 2007).

The determination of sequence variation on nucleotide or short subsequence level is also possible without genome assemblies. This is especially important for quasispecies assemblies with high heterogeneity, for which sequence assembly from short reads is fundamentally limited. In such cases, virus sequence reads need to be aligned against a reference genome sequence. Many variant detection tools can extract variants from these alignments, even for low abundant variants that only occur in a small portion of reads (Koboldt et al. 2009).

1.3 PROTEIN FAMILIES AND ORTHOLOGOUS GROUPS OF VIRUSES

Due to the rapid evolution of virus genome sequences, direct comparisons of genomic sequences are only possible for closely related genomes. Comparisons across larger evolutionary distances require an alternative approach, which makes use of evolutionary constraints that apply to conserved loci within the genomes. Most popular for this purpose are protein-coding genes, which translate into amino acid sequences. Although structures are even more conserved than protein sequences, fundamental limitations in the prediction of protein structures from sequence data apply. For this reason, protein sequence comparisons are to-date the most powerful and universal approach to comprehensive comparative genomics of viruses.

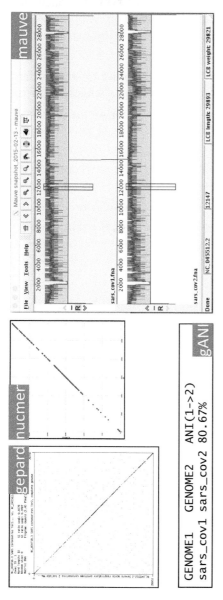

FIGURE 1.2 Demonstration of the comparison of the same two genomes with Mauve, nucmer (Mummer), gepard, gANI, and show-snps (Mummer).

In principle, comparative genomics of viral protein sequences is based on the same toolbox of sequence bioinformatics as it is used for cellular genomes. In brief, protein sequences are clustered into homologous or orthologous groups (El-Gebali et al. 2019). For each group, multiple alignments are calculated and converted into position-specific conservation models. These models can be used to recruit further protein sequences, as applied for automatic annotation of protein sequences, and to analyze the presence and absence or group members in genomes, which is frequently used to compare genomes. However, the main challenges for virus genomes are the low sequence conservation of virus proteins, compared to many proteins in cellular genomes, and the presence of polyproteins, in which multiple proteins are encoded in the same open reading frame. Whereas the low sequence conservation can be partially compensated by comparisons of 3D protein structures (Figure 1.3), no consistent approach exists for the automatic annotation of polyproteins (Figure 1.4).

The comparative analysis of virus genomes, based on their protein sequences, is so far impaired by the inhomogeneity of virus genome annotations in the public INSDC databases (Cochrane et al. 2016) as well as in NCBI

FIGURE 1.3 Example of a structurally conserved sequence singleton: PRD1 capsid P3 in xfam.org family PF09018.

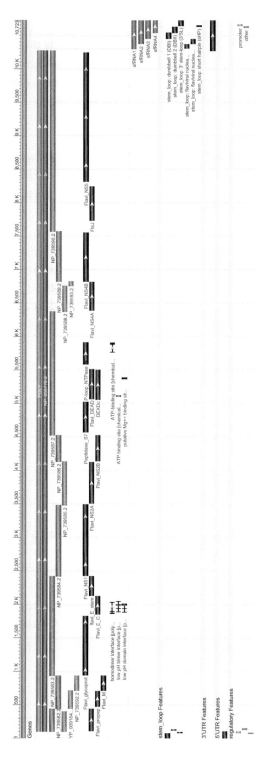

FIGURE 1.4 Example of a typical polyprotein: NP_056776.2 polyprotein [Dengue virus 2], which contains structural proteins, C-prM/M-E and non-structural proteins, NS1-NS2A/B-NS3-NS4A/B-NS5.

RefSeq (O'Leary et al. 2016). Even very closely related virus sequences can potentially differ for technical reasons, such as the use of different gene prediction software or different reference databases for homology-based annotation transfer. This problem can so far only be solved by consistent re-annotation of genomes prior to the calculation of protein homologous or orthologous groups (Schaffer et al. 2020). Secondary databases, which provide reannotations of all virus genome sequences, often lack long-term support (e.g., Firth 2014).

Public resources for protein families can be grouped into generic and virus-specific databases. The most important generic databases for protein families are covered by Interpro (Mitchell et al. 2019). This project groups entries from multiple member databases into one consistent system of protein domains and protein families, and also provides server-based and stand-alone search tools for sequence annotation and comparison. Virus-specific orthologous groups are provided by the eggNOG database, which includes thousands of well-annotated virus genomes (Huerta-Cepas et al. 2019). The Virus Pathogen Resource VIPR provides pre-calculated orthologous groups for families of viral pathogens (Pickett et al. 2012). The Prokaryotic Virus Orthologous Groups (pVOGs) contain orthologous groups of phage proteins (Grazziotin, Koonin, and Kristensen 2017). The recently developed VOGDB (http://vogdb.org) contains all virus genomes from NCBI RefSeq and group their protein sequences into protein families in a multistep approach that is based on orthology and remote homology. VOGDB families are classified into functional categories and are annotated according to their conservation in cellular genomes.

1.4 EVOLUTION OF PROTEIN FAMILIES WITHIN VIRUS AND HOST GENOMES

Viral genomes and their encoded proteins are shaped by the evolution of the viruses and their hosts. Protein families, e.g., can be crucial for the replication of a virus, can be targeting the host defense, or can be beneficial for the host's fitness. Early studies on comparative genomics of viruses suggested the grouping of protein families into several categories, according to their genomic occurrences. Five classes that can be assorted into three larger categories seemed to be well-distinguishable (Koonin, Senkevich, and Dolja 2006):

1. Genes with readily detectable homologs in cellular life forms
   a) Genes with closely related homologs in cellular organisms (typically, the host of the given virus) present in a narrow group of viruses.

b) Genes that are conserved within a major group of viruses or even several groups and have relatively distant cellular homologs.

2. Virus-specific genes

c) ORFans, i.e., genes without detectable homologs except, possibly, in closely related viruses.

d) Virus-specific genes that are conserved in a (relatively) broad group of viruses but have no detectable homologs in cellular life forms

3. Viral hallmark genes

e) Genes shared by many diverse groups of viruses, with only distant homologs in cellular organisms, and with strong indications of monophyly of all viral members of the respective gene families—we would like to coin the phrase "viral hallmark genes" to denote these genes that can be viewed as distinguishing characters of the "virus state."

In the all-virus collection of virus protein families VOGDB, still the vast majority of groups is virus specific—irrespective of the stringency criteria applied (Table 1.1). Most of these virus-specific families, however, are small and thus can be considered ORFans. The largest virus-specific groups contain up to 300 hypothetical proteins from up to 150 different genomes.

1.5 OUTLOOK

Despite the progress in virus comparative genomics during the last years, this field will need massive improvements in the near future. High throughput sequencing of long reads, combined with powerful error-correction algorithms, will lead to massively increased numbers of high-quality virus genome sequences, which need to be organized and compared. Of most importance are powerful, fully automatic genome annotation workflows that allow the re-annotation of sequence submissions in public databases.

TABLE 1.1   Number of Total Proteins and Virus-Specific Proteins in VOGDB (http://vogdb.org) Release 201

Total number of protein families	25,743
Virus specific (high stringency)	20,358
Virus specific (medium stringency)	21,986
Virus specific (low stringency)	22,703

Criteria for high stringency: no hits of E<1e−04 in more than two cellular genomes; medium stringency: no hits of E<1e−10 in more than three cellular genomes; low stringency: no hits of E<1e−15 in more than four cellular genomes.

This concept has already been very successful for prokaryotic genomes (Haft et al. 2018) and needs to be extended to viruses. Further downstream, automatic grouping of viral strains and variants into pangenomes will help to control the computational costs of the calculation of homologous and orthologous groups. For this purpose, suitable concepts from prokaryotic comparative genomes could be adapted to viruses in the future.

REFERENCES

Cochrane, G., I. Karsch-Mizrachi, T. Takagi, and Collaboration International Nucleotide Sequence Database. 2016. "The international nucleotide sequence database collaboration." *Nucleic Acids Res* 44 (D1):D48–50. doi: 10.1093/nar/gkv1323.

Darling, A. C., B. Mau, F. R. Blattner, and N. T. Perna. 2004. "Mauve: Multiple alignment of conserved genomic sequence with rearrangements." *Genome Res* 14 (7):1394–403. doi: 10.1101/gr.2289704.

El-Gebali, S., J. Mistry, A. Bateman, S. R. Eddy, A. Luciani, S. C. Potter, M. Qureshi, L. J. Richardson, G. A. Salazar, A. Smart, E. L. L. Sonnhammer, L. Hirsh, L. Paladin, D. Piovesan, S. C. E. Tosatto, and R. D. Finn. 2019. "The Pfam protein families database in 2019." *Nucleic Acids Res* 47 (D1):D427–32. doi: 10.1093/nar/gky995.

Firth, A. E. 2014. "Mapping overlapping functional elements embedded within the protein-coding regions of RNA viruses." *Nucleic Acids Res* 42 (20):12425–39. doi: 10.1093/nar/gku981.

Gifford, R. J., J. Blomberg, J. M. Coffin, H. Fan, T. Heidmann, J. Mayer, J. Stoye, M. Tristem, and W. E. Johnson. 2018. "Nomenclature for endogenous retrovirus (ERV) loci." *Retrovirology* 15 (1):59. doi: 10.1186/s12977-018-0442-1.

Grazziotin, A. L., E. V. Koonin, and D. M. Kristensen. 2017. "Prokaryotic Virus Orthologous Groups (pVOGs): A resource for comparative genomics and protein family annotation." *Nucleic Acids Res* 45 (D1):D491–8. doi: 10.1093/nar/gkw975.

Greninger, A. L. 2018. "A decade of RNA virus metagenomics is (not) enough." *Virus Res* 244:218–29. doi: 10.1016/j.virusres.2017.10.014.

Haft, D. H., M. DiCuccio, A. Badretdin, V. Brover, V. Chetvernin, K. O'Neill, W. Li, F. Chitsaz, M. K. Derbyshire, N. R. Gonzales, M. Gwadz, F. Lu, G. H. Marchler, J. S. Song, N. Thanki, R. A. Yamashita, C. Zheng, F. Thibaud-Nissen, L. Y. Geer, A. Marchler-Bauer, and K. D. Pruitt. 2018. "RefSeq: An update on prokaryotic genome annotation and curation." *Nucleic Acids Res* 46 (D1):D851–60. doi: 10.1093/nar/gkx1068.

Huerta-Cepas, J., D. Szklarczyk, D. Heller, A. Hernandez-Plaza, S. K. Forslund, H. Cook, D. R. Mende, I. Letunic, T. Rattei, L. J. Jensen, C. von Mering, and P. Bork. 2019. "eggNOG 5.0: A hierarchical, functionally and phylogenetically annotated orthology resource based on 5090 organisms and 2502 viruses." *Nucleic Acids Res* 47 (D1):D309–14. doi: 10.1093/nar/gky1085.

Koboldt, D. C., K. Chen, T. Wylie, D. E. Larson, M. D. McLellan, E. R. Mardis, G. M. Weinstock, R. K. Wilson, and L. Ding. 2009. "VarScan: Variant detection in massively parallel sequencing of individual and pooled samples." *Bioinformatics* 25 (17):2283–5. doi: 10.1093/bioinformatics/btp373.

Koonin, E. V., T. G. Senkevich, and V. V. Dolja. 2006. "The ancient Virus World and evolution of cells." *Biol Direct* 1:29. doi: 10.1186/1745-6150-1-29.

Krumsiek, J., R. Arnold, and T. Rattei. 2007. "Gepard: A rapid and sensitive tool for creating dotplots on genome scale." *Bioinformatics* 23 (8):1026–8. doi: 10.1093/bioinformatics/btm039.

Libin, P. J. K., K. Deforche, A. B. Abecasis, and K. Theys. 2019. "VIRULIGN: Fast codon-correct alignment and annotation of viral genomes." *Bioinformatics* 35 (10):1763–5. doi: 10.1093/bioinformatics/bty851.

Marcais, G., A. L. Delcher, A. M. Phillippy, R. Coston, S. L. Salzberg, and A. Zimin. 2018. "MUMmer4: A fast and versatile genome alignment system." *PLoS Comput Biol* 14 (1):e1005944. doi: 10.1371/journal.pcbi.1005944.

Masson, P., C. Hulo, E. De Castro, H. Bitter, L. Gruenbaum, L. Essioux, L. Bougueleret, I. Xenarios, and P. Le Mercier. 2013. "ViralZone: Recent updates to the virus knowledge resource." *Nucleic Acids Res* 41 (Database issue):D579–83. doi: 10.1093/nar/gks1220.

Mitchell, A. L., T. K. Attwood, P. C. Babbitt, M. Blum, P. Bork, A. Bridge, S. D. Brown, H. Y. Chang, S. El-Gebali, M. I. Fraser, J. Gough, D. R. Haft, H. Huang, I. Letunic, R. Lopez, A. Luciani, F. Madeira, A. Marchler-Bauer, H. Mi, D. A. Natale, M. Necci, G. Nuka, C. Orengo, A. P. Pandurangan, T. Paysan-Lafosse, S. Pesseat, S. C. Potter, M. A. Qureshi, N. D. Rawlings, N. Redaschi, L. J. Richardson, C. Rivoire, G. A. Salazar, A. Sangrador-Vegas, C. J. A. Sigrist, I. Sillitoe, G. G. Sutton, N. Thanki, P. D. Thomas, S. C. E. Tosatto, S. Y. Yong, and R. D. Finn. 2019. "InterPro in 2019: Improving coverage, classification and access to protein sequence annotations." *Nucleic Acids Res* 47 (D1):D351–60. doi: 10.1093/nar/gky1100.

O'Leary, N. A., M. W. Wright, J. R. Brister, S. Ciufo, D. Haddad, R. McVeigh, B. Rajput, B. Robbertse, B. Smith-White, D. Ako-Adjei, A. Astashyn, A. Badretdin, Y. Bao, O. Blinkova, V. Brover, V. Chetvernin, J. Choi, E. Cox, O. Ermolaeva, C. M. Farrell, T. Goldfarb, T. Gupta, D. Haft, E. Hatcher, W. Hlavina, V. S. Joardar, V. K. Kodali, W. Li, D. Maglott, P. Masterson, K. M. McGarvey, M. R. Murphy, K. O'Neill, S. Pujar, S. H. Rangwala, D. Rausch, L. D. Riddick, C. Schoch, A. Shkeda, S. S. Storz, H. Sun, F. Thibaud-Nissen, I. Tolstoy, R. E. Tully, A. R. Vatsan, C. Wallin, D. Webb, W. Wu, M. J. Landrum, A. Kimchi, T. Tatusova, M. DiCuccio, P. Kitts, T. D. Murphy, and K. D. Pruitt. 2016. "Reference sequence (RefSeq) database at NCBI: Current status, taxonomic expansion, and functional annotation." *Nucleic Acids Res* 44 (D1):D733–45. doi: 10.1093/nar/gkv1189.

Olm, M. R., C. T. Brown, B. Brooks, and J. F. Banfield. 2017. "dRep: A tool for fast and accurate genomic comparisons that enables improved genome recovery from metagenomes through de-replication." *ISME J* 11 (12):2864–8. doi: 10.1038/ismej.2017.126.

Pickett, B. E., E. L. Sadat, Y. Zhang, J. M. Noronha, R. B. Squires, V. Hunt, M. Liu, S. Kumar, S. Zaremba, Z. Gu, L. Zhou, C. N. Larson, J. Dietrich, E. B. Klem, and R. H. Scheuermann. 2012. "ViPR: An open bioinformatics database and analysis resource for virology research." *Nucleic Acids Res* 40 (Database issue):D593–8. doi: 10.1093/nar/gkr859.

Roux, S., F. Enault, B. L. Hurwitz, and M. B. Sullivan. 2015. "VirSorter: Mining viral signal from microbial genomic data." *PeerJ* 3:e985. doi: 10.7717/peerj.985.

Roux, S., E. M. Adriaenssens, B. E. Dutilh, E. V. Koonin, A. M. Kropinski, M. Krupovic, J. H. Kuhn, R. Lavigne, J. R. Brister, A. Varsani, C. Amid, R. K. Aziz, S. R. Bordenstein, P. Bork, M. Breitbart, G. R. Cochrane, R. A. Daly, C. Desnues, M. B. Duhaime, J. B. Emerson, F. Enault, J. A. Fuhrman, P. Hingamp, P. Hugenholtz, B. L. Hurwitz, N. N. Ivanova, J. M. Labonte, K. B. Lee, R. R. Malmstrom, M. Martinez-Garcia, I. K. Mizrachi, H. Ogata, D. Paez-Espino, M. A. Petit, C. Putonti, T. Rattei, A. Reyes, F. Rodriguez-Valera, K. Rosario, L. Schriml, F. Schulz, G. F. Steward, M. B. Sullivan, S. Sunagawa, C. A. Suttle, B. Temperton, S. G. Tringe, R. V. Thurber, N. S. Webster, K. L. Whiteson, S. W. Wilhelm, K. E. Wommack, T. Woyke, K. C. Wrighton, P. Yilmaz, T. Yoshida, M. J. Young, N. Yutin, L. Z. Allen, N. C. Kyrpides, and E. A. Eloe-Fadrosh. 2019. "Minimum Information about an Uncultivated Virus Genome (MIUViG)." *Nat Biotechnol* 37 (1):29–37. doi: 10.1038/nbt.4306.

Schaffer, A. A., E. L. Hatcher, L. Yankie, L. Shonkwiler, J. R. Brister, I. Karsch-Mizrachi, and E. P. Nawrocki. 2020. "VADR: Validation and annotation of virus sequence submissions to GenBank." *BMC Bioinf* 21 (1):211. doi: 10.1186/s12859-020-3537-3.

Schulz, F., S. Roux, D. Paez-Espino, S. Jungbluth, D. A. Walsh, V. J. Denef, K. D. McMahon, K. T. Konstantinidis, E. A. Eloe-Fadrosh, N. C. Kyrpides, and T. Woyke. 2020. "Giant virus diversity and host interactions through global metagenomics." *Nature* 578 (7795):432–6. doi: 10.1038/s41586-020-1957-x.

Schuster, P. 2016. "Quasispecies on fitness landscapes." *Curr Top Microbiol Immunol* 392:61–120. doi: 10.1007/82_2015_469.

Shean, R. C., N. Makhsous, G. D. Stoddard, M. J. Lin, and A. L. Greninger. 2019. "VAPiD: A lightweight cross-platform viral annotation pipeline and identification tool to facilitate virus genome submissions to NCBI GenBank." *BMC Bioinf* 20 (1):48. doi: 10.1186/s12859-019-2606-y.

Swetina, J., and P. Schuster. 1982. "Self-replication with errors: A model for polynucleotide replication." *Biophys Chem* 16 (4):329–45. doi: 10.1016/0301-4622(82)87037-3.

Thannesberger, J., H. J. Hellinger, I. Klymiuk, M. T. Kastner, F. J. J. Rieder, M. Schneider, S. Fister, T. Lion, K. Kosulin, J. Laengle, M. Bergmann, T. Rattei, and C. Steininger. 2017. "Viruses comprise an extensive pool of mobile genetic elements in eukaryote cell cultures and human clinical samples." *FASEB J* 31 (5):1987–2000. doi: 10.1096/fj.201601168R.

Topfer, A., T. Marschall, R. A. Bull, F. Luciani, A. Schonhuth, and N. Beerenwinkel. 2014. "Viral quasispecies assembly via maximal clique enumeration." *PLoS Comput Biol* 10 (3):e1003515. doi: 10.1371/journal.pcbi.1003515.

UniProt Consortium. 2019. "UniProt: A worldwide hub of protein knowledge." *Nucleic Acids Res* 47 (D1):D506–15. doi: 10.1093/nar/gky1049.

Varghese, N. J., S. Mukherjee, N. Ivanova, K. T. Konstantinidis, K. Mavrommatis, N. C. Kyrpides, and A. Pati. 2015. "Microbial species delineation using whole genome sequences." *Nucleic Acids Res* 43 (14):6761–71. doi: 10.1093/nar/gkv657.

Yamashita, T., H. Takeda, A. Takai, S. Arasawa, F. Nakamura, Y. Mashimo, M. Hozan, S. Ohtsuru, H. Seno, Y. Ueda, and A. Sekine. 2020. "Single-molecular real-time deep sequencing reveals the dynamics of multi-drug resistant haplotypes and structural variations in the hepatitis C virus genome." *Sci Rep* 10 (1):2651. doi: 10.1038/s41598-020-59397-2.

CHAPTER 2

Current Techniques and Approaches for Metagenomic Exploration of Phage Diversity

Simon Roux

Lawrence Berkeley National Laboratory

Mark Borodovsky

Georgia Tech

CONTENTS

2.1 INTRODUCTION

Viruses are present in every ecosystem on Earth, and often vastly outnumber their cellular hosts. While current surveys are far from exhaustive, it is already clear that global viral diversity is extremely large and complex. Among this virosphere, bacteriophages—or phages, i.e., viruses infecting bacteria—represent one of the most commonly identified types of virus, consistent with their host representing one of the most abundant forms of life. The global phage diversity itself is very broad, including both RNA and DNA genomes, with size ranging from a few kilobases to half a megabase, and a high proportion of genes unique to each individual phage and not detected in any other genome (Dion, Oechslin, and Moineau 2020). Across earth's biomes, phages constantly infect, alter, reprogram, and kill their bacterial host cell, leading to substantial shifts in microbial community structure, function, and metabolic output (Shkoporov and Hill 2019; Williamson et al. 2017; Breitbart et al. 2018; Trubl et al. 2020). A full understanding of microbiome processes and drivers can thus not be achieved without a thorough characterization of its resident phages and their associated impacts.

The broad diversity and importance of viruses, especially bacteriophages, was only recently recognized. Historically, microbial communities, including viruses, were mostly explored through laboratory cultivation, which strongly underestimated natural diversity. While PCR-based exploration of microbes using, e.g., 16S amplicons quickly expanded our collective knowledge of microbial life on Earth, the same approaches could not be applied to viruses because they lack a single universal marker gene. Hence, it's only with the advent of high-throughput metagenomics, i.e., whole-genome shotgun sequencing applied to genetic material directly extracted from the environment, that the full extent of viral diversity was progressively revealed (Sullivan 2015). Since then, however, metagenomics quickly established itself as the method of choice for exploring phage diversity, leading to an exponential increase in the number of phage genomes in databases and several paradigm shifts about their role in nature.

2.2 PHAGE METAGENOMICS: A BRIEF HISTORY

When pioneered in the early 2000s, viral metagenomes (viromes) consisted of a collection of a few thousands short gene fragments and provided an unprecedented yet very limited view of complex environmental viral communities. These first datasets all pointed toward the existence of an extensive phage diversity (Edwards and Rohwer 2005), but the fragmented nature of the data greatly limited the type and scope of analyses applicable (Brum and Sullivan 2015). These limitations were progressively overcome through a combined improvement in sequencing throughput and in tools used for metagenome assembly, so that by the years ~2015, complete or near-complete phage genomes could be routinely reconstructed from short metagenomic fragments. This process of genome assembly for metagenomes was especially efficient for phages due to their relatively short genome size and low frequency of exact repeats compared to their bacterial and archaeal hosts.

Since then, metagenomics has transformed the way scientists can identify and study viruses in the environment (Hayes et al. 2017; Brum and Sullivan 2015; Sutton and Hill 2019; Trubl et al. 2020). This is illustrated by the quick rise of virus genomes and genome fragments assembled from metagenomes and deposited into public databases (Figure 2.1a): in 2010, only ~100 viral genomes (and genome fragments) assembled from metagenomes were publicly available, while this number reached 35,000 in 2016, 775,000 in 2018, and 2,300,000 sequences in 2020 (Páez-Espino et al. 2016, 2018). Phage genome exploration via metagenomics has been successfully performed in a broad range of ecosystems (Aggarwala, Liang, and Bushman 2017; Anantharaman et al. 2014; Gregory et al. 2019; Emerson et al. 2018), and includes both DNA and RNA genomes (Callanan et al. 2018). As a growing number of researchers and studies used metagenomics to explore and characterize uncultivated phages, a common framework was developed for the generation and analysis of these sequences to share best practices and enable comparative analyses (Roux et al. 2019). Exploration of phage communities via metagenomics typically includes six steps: metagenome assembly, phage sequence identification and quality assessment, functional annotation, taxonomic classification, host taxonomy prediction, and estimation of distribution and prevalence (Figure 2.1b). In this pipeline, the metagenome assembly and phage sequence identification steps are especially critical as all other analyses will be applied to their products, i.e., the set of phage genomes (and genome fragments) extracted from the assembled data.

2.3 RECOVERING PHAGE GENOMES FROM METAGENOMES

Metagenome assemblies for phage exploration are most often obtained from standard tools and pipelines used for general short-read metagenomics. Since the vast majority of phage diversity is novel compared to current databases, *de novo* assembly approaches have typically been favored over

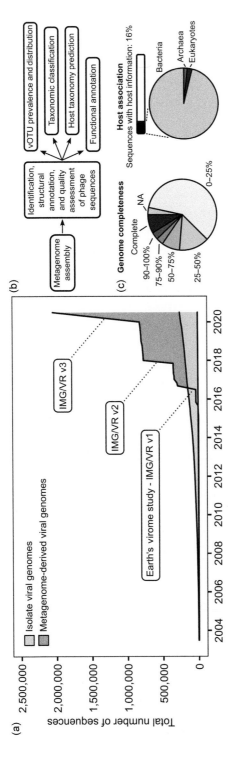

FIGURE 2.1   Growth, processing overview, and main characteristics of the IMG/VR database of metagenome-derived viral sequences (https://img.jgi.doe.gov/vr/). (a) Growth of the number of metagenome-derived viral sequences in comparison to the number of viral genomes from isolates available in NCBI Viral RefSeq. The vast majority of these metagenome-derived viral sequences are available in the IMG/VR database, from which three major releases are indicated on the plot. (b) Current analysis framework for metagenome-derived viral sequences. (c) Characteristics of the IMG/VR 3.0 (August 2020) database including genome completeness evaluated with CheckV (left) and domain-level host association based on CRISPR matches, similarities to known viruses, and similarities to host genomes (right).

reference-based assembly, although the latter is frequently used in clinical metagenomics, e.g., to assemble new variants of a known viral pathogen (Fancello, Raoult, and Desnues 2012). Empirically, researchers in the field have observed that tools initially developed for the *de novo* assembly of microbial genomes, such as Idba_ud (Peng et al. 2012), Megahit (Li et al. 2016), or MetaSPAdes (Nurk et al. 2017), were very efficient at assembling phage genomes as well, and there was little value in developing alternative approaches. For instance, metaViralSPAdes was recently designed specifically for the purpose of assembling and identifying viral genomes from metagenomes; however, this tool uses a standard approach for metagenome assembly, and the viral-specific part is mostly associated with the identification of viral contigs (see below).

Exploration of microbial diversity through metagenomics now routinely includes a genome binning step after metagenome assembly to identify which sequences derived from the same original genome (Bowers et al. 2017). While in theory a similar approach could be used for phages (and has been in some cases, e.g. (Roux et al. 2016)), in practice it is often omitted in phage metagenomics pipelines. As most phage genomes are shorter than 100kb, they are often assembled in a single large piece, and genome binning does not provide a significant improvement in genome recovery. Since even "jumbo" phages, with genomes larger than 300 kb, can now be entirely assembled from metagenomes (Devoto et al. 2019; Al-Shayeb et al. 2020), it is likely that most viral genomes obtained from metagenomes in the next few years will be directly assembled from metagenome reads without a genome binning step.

In addition to genome recovery from short-read metagenome assemblies, long-reads metagenomics, i.e., metagenomes sequenced with PacBio or ONT technologies often yielding reads of 10kb and more, are quickly becoming a viable approach for phage exploration. Long-reads metagenomics holds fantastic promises for phage diversity exploration, including the recovery of hypervariable regions of phage genomes which are often not assembled from short reads (Warwick-Dugdale et al. 2019). Phages genomes could also be sequenced as single (long) reads, enabling large-scale studies of population variation and recombination that remain challenging to perform with short-reads metagenomes, although such studies will require a higher base-calling accuracy than current technologies provide. As of now, long-reads metagenomics for phage exploration is still in its infancy and must overcome a number of technological barriers such as how to extract long and intact genome fragments from environmental viral communities, but long-reads technologies will almost certainly play an increasingly large role in the discovery of new phage genomes through metagenomics.

2.4 IDENTIFICATION AND QUALITY CONTROL OF PHAGE CONTIGS IN METAGENOME ASSEMBLY

Two main approaches have been proposed to identify known and novel phage genomes in an assembled metagenome, with variations around these two approaches declined across multiple tools (e.g., Arndt et al. 2016; Roux et al. 2015; Amgarten et al. 2018; Ren et al. 2017, 2020; Kieft, Zhou, and Anantharaman 2020). The first approach relies on predicted genome features, primarily gene content, and has been originally built on previous efforts to identify phages and other mobile genetic elements in whole-genome shotgun sequencing datasets (Lima-Mendez et al. 2008; Akhter, Aziz, and Edwards 2012). These identification methods leverage several distinctive features of phage genomes that have been identified from isolates, including (i) a high proportion of viral assignation among genes with a confident taxonomic annotation, especially genes uniquely found on phage genomes such as the ones coding for elements of the viral capsid; (ii) a high proportion of "novel" genes, i.e., genes without any similarity to known functional domains; and (iii) a tendency to encode genes in long contiguous stretches on the same strand. Enrichment statistics and empirical thresholds were often used in early versions of these tools, while the more recent ones now use machine learning algorithms, in particular random forest classifiers, to distinguish viral from cellular sequences (Kieft, Zhou, and Anantharaman 2020; Amgarten et al. 2018). This type of approach has consistently proven to be efficient and reliable, with two main caveats. First, other mobile genetic elements such as some plasmids can share some of these characteristics (especially the lack of functional annotation) and wrongly get identified as "viral." Second, their efficiency is highly dependent on the size of the input contig: for short contigs encoding only a few genes, the signal used by these approaches is usually not strong enough to yield a confident prediction, leading to a high proportion of false negatives, i.e., short viral contigs that remain undetected (Roux et al. 2015).

The second type of approach for the detection of phage contigs in metagenome assemblies uses nucleotide composition (in practice frequency of k-mers) paired with a machine (deep-)learning approach and large databases of known viral and nonviral genomes (Ren et al. 2017, 2020). These methods do not require the input contig to be functionally annotated, and thus are not impacted by potential errors in gene prediction such as overlapping genes or programmed frameshifts (see below), and can be efficient on short contigs (~1 kb) which typically include enough signal in terms of nucleotide composition (Ren et al. 2017). There are however several practical limitations and challenges when using k-mer-based approaches, which are mainly associated with the composition and diversity of the training databases. Based on current benchmarks, it seems like these methods are not as efficient as the ones based on gene content for the identification of entirely novel phages, i.e., phages without any close relative in the existing databases. In addition, nonviral sequences including plasmids and fragments of eukaryotic genomes can be identified as "viral" with a very high confidence using k-mer based approaches, while the same fragments would be correctly identified as "nonviral" by gene-content approaches (Ponsero and Hurwitz 2019). This is likely due to the fact that, given the large diversity of viral sequences, some of these methods really "learn" to recognize microbial genomes and classify everything else as "confident" viral sequences. While these limitations will be progressively overcome, especially via the increased phage diversity in databases, they need to be kept in mind when analyzing sequences predicted as "viral" by k-mer based tools.

Finally, a quality estimation of the genome is often performed along with the contig identification (Figure 2.1b and c). These estimations are typically based on an estimation of genome completeness based on comparison to a database of (predicted) complete genomes, as well as the identification of putative host regions on a contig initially detected as mainly viral. The tool VIBRANT (Kieft, Zhou, and Anantharaman 2020), for instance, includes both an identification and quality assessment module assigning contigs to quality tiers. Meanwhile, CheckV (Nayfach et al. 2020) is a tool entirely dedicated to quality evaluation, including quantitative estimation of genome completeness. While these tools are still very recent, it is quickly becoming a "good" practice to use them immediately after phage contig identification, as they provide a useful overview of the quality of the dataset extracted. In the large IMG/VR database for instance, mostly based on short-reads metagenomes, a majority of the sequences represent short genomic fragments, and a quality assessment tool like CheckV enables an automated detection of the subset of sequences representing complete or near-complete genomes (Figure 2.1c).

2.5 STRUCTURAL ANNOTATION OF METAGENOME-ASSEMBLED PHAGE GENOMES

The task of gene prediction in phage genomes presents a challenge for conventional whole genome prokaryotic gene finders such as Glimmer or Prodigal (Delcher et al. 2007; Hyatt et al. 2010) due to several phage-specific features: (i) a short genome length, (ii) more frequent overlapping genes, (iii) possible programmed frameshifts, and (iv) unknown genetic code. For metagenome-assembled phage genomes, the short genome length is an especially serious obstacle; a phage genome assembly could even consist of several pieces to be analyzed independently if they are not presented by an assembly algorithm as parts of a scaffold. Importantly, the nature of a sequence fragment, whether it is a sequence from a phage or from a microbe, may be revealed only at the stage of functional annotation (see above). Fortunately, the task of gene prediction in fragments of phage genomes is very similar to the task of gene prediction in anonymous fragments of prokaryotic genomes that appear in metagenomes. Therefore, we will focus here on the gene finders that predict genes in metagenomes, such as Glimmer-MG, MetaProdigal, and others (Kelley et al. 2012; Hyatt et al. 2012; Rho, Tang, and Ye 2010; Noguchi, Taniguchi, and Itoh 2008), including the recently developed GeneMarkS-2 (Lomsadze et al. 2018) that has special modes of operation for both long and short genomic contigs. If a contig contains <100 genes, GeneMarkS-2 does not proceed with unsupervised estimation of model parameters performed by iterative self-training. Instead, the algorithm automatically switches to MetaGeneMark-2 (Lomsadze et al. 2018) that determines parameters of high order models with help of analytical approximations of oligomer frequencies as functions of the genome GC content (Zhu et al. 2010; Besemer and Borodovsky 1999). For prokaryotic genomic sequences GeneMarkS-2 was shown to possess the highest accuracy both in the detection of a gene as a whole and in exact pinpointing the gene start (Lomsadze et al. 2018). For assembled contigs with > 100 genes, GeneMarkS-2 constructs a model of regulatory signals situated in gene upstream regions, such as ribosome binding site (RBS). Interestingly, it was observed that the RBS nucleotide frequency pattern derived by GeneMarkS-2 for a phage was similar to the one of its host genome, whether it was the Shine–Dalgarno pattern that appears in more than 50% of prokaryotic genomes, or less frequent, non-canonical, AT-rich RBS pattern. The self-training modeling capability of GeneMarkS-2, could also infer sequence patterns in gene upstream regions associated with leaderless transcription. All over, the self-training for models of signals near gene starts did increase the accuracy of gene start prediction (Lomsadze et al. 2018).

Protein-coding gene overlaps in genomic sequences were first discovered in the phage PhiX174 at the end of 1970s. The partial unidirectional gene overlaps in phages are frequent, and they carry an evolutionary advantage by enabling reduction of the size of the phage genome. Some phages have overlaps over the whole length of one of the genes, e.g., PhiX174, where four of ten genes are overlapped by others. Prediction of a gene overlap that covers a whole gene or its large part (>300 nt) is a difficult task for ab initio gene finders. Still, only a few experimentally confirmed cases of such long gene overlaps have been known. Short gene overlaps (<300 nt) are predicted by most gene finders, particularly by GeneMarkS-2.

Unidirectional gene overlaps have to be discriminated from frameshifts. Some frameshifts observed in phage genomic sequences could be artifacts caused by sequencing errors, while others, the programmed frameshifts, have a regulatory role and are conserved across evolution. Interestingly, programmed frameshifts were observed more frequently in phages than in their hosts (Antonov et al. 2013; Baranov et al. 2006). Specialized tools, such as GeneTack (Antonov and Borodovsky 2010), have been developed to predict frameshifts. There are several genomic sequence determinants that help carry out this task. Short unidirectional gene overlaps would have a stop codon of the upstream gene close to the position of the frameshift, with gene overlaps of length one and four especially frequent. Programmed frameshifts, on the other hand, would have a so-called frameshift signal in close proximity to the frameshift position, either a "slippery k-mer" associated with +1 frameshifts or an "internal RBS site" to slow down the ribosome in case of −1 frameshifts (Antonov and Borodovsky 2010).

Accurate gene prediction in phage contigs reconstructed from metagenomes requires knowledge of the sequence genetic code. In particular, the TGA triplet could be a stop codon (Genetic code 11) or a codon for Trp (Genetic code 4) or a codon for Gly (Genetic code 25). Recognition of a type of genetic code is challenging for short phage contigs. The latest version of MetaGeneMark-2 can recognize the type of genetic code of a phage contig, assuming that the code is uniform across the sequence. However, in some phages, parts of genomic sequence could contain genes translated in similar ways by two types of genetic code, thus presenting an additional challenge that, so far, no automated tool can efficiently address (Ivanova et al. 2014; Yutin et al. 2020). Overall, several existing gene finders could be successfully applied for gene prediction in the metagenome-assembled phage contigs. Such reference-free gene finders (that do not need an expert prepared training set) use robust statistical models and produce sufficiently high accuracy of prediction for a large fraction of genes. Meanwhile, prediction of large gene overlaps as well as of very short genes, though infrequent, remains a challenging task.

2.6 BRINGING MEANINGFUL ECO-EVOLUTIONARY CONTEXT TO METAGENOME-ASSEMBLED PHAGE GENOMES

Following assembly, identification, quality assessment, and structural annotation, metagenome-assembled phage genomes can be used in a broad range of analyses depending on the type of dataset and research question addressed. Some of the most common include functional annotation, taxonomic classification, and in silico host prediction.

Functional annotation of cds predicted on phage genomes typically relies on remote homology searches based on HMM profiles (Roux et al. 2019), which enable the identification of conserved domains even for phage sequences which tend to be highly divergent from well-characterized references. A combination of generalist databases—such as Pfam (Finn et al. 2016), CDD (Marchler-Bauer et al. 2015), and KEGG (Kanehisa et al. 2016)—along with viral-dedicated databases—such as pVOGs (Grazziotin, Koonin, and Kristensen 2017) and VOGdb (http://vogdb.org)—is typically used in an attempt to maximize the number of annotated genes. Nevertheless, in current studies, a majority of cds remain entirely uncharacterized even after this functional annotation step, highlighting the need for further targeted experimental characterization of phage-encoding functional diversity.

A number of tools now enable genome-based taxonomic classification for viruses (Simmonds et al. 2017). Achieving robust taxonomic classification is still difficult for partial genomes, however, especially when derived from phages with no close relatives in the databases. For ecological studies, phage genomes assembled from metagenomes are often clustered into species-level groups, also known as "vOTU" (Roux et al. 2019), based on average nucleotide identity. Reads from several metagenomes are then typically mapped to representative sequences of these vOTUs, which enables the analysis of distribution of these phages across samples (based on presence/absence), as well as prevalence (using read coverage as a proxy for relative abundance) if analyzing a quantitative metagenome (Roux et al. 2019). In addition, the same read mapping results can be used for population genetics studies of uncultivated phages, although no automatic tool is yet available for this type of analysis (Ignacio-espinoza, Ahlgren, and Fuhrman, n.d.; Gregory et al. 2019).

Most studies of phage genomes from metagenomes also attempt to link new phages to a putative host. Different methods have been proposed for this purpose (Edwards et al. 2016), primarily based on (i) large (>1 kb) sequence similarity between phage and host genomes, (ii) sequence similarity between a phage genome and predicted CRISPR spacer(s) (Zhang et al. 2020), and (iii) similarity between phage and host genome in terms of nucleotide composition (Ahlgren et al. 2016; Galiez et al. 2017). Some of the most recent tools also attempt to predict virus-host pairs by integrating several of these methods (Wang et al. 2019). All these tools are, however, still relatively new, and their results must be cautiously evaluated. Notably, these tools are highly dependent on the host reference database, and the results are often improved when users can supplement existing databases with newly assembled microbial genomes from the same environment from which the phage genomes were obtained (Roux et al. 2016). Hence, linking newly discovered phages to their hosts is highly desirable but remains a challenging process requiring multiple rounds of manual inspection and curation at the moment. Finally, the last analysis frequently performed on new phage genomes is to mine the functional annotation for "Auxiliary Metabolic Genes" (AMGs), i.e., genes used during the infection to alter host cell metabolism (Breitbart et al. 2018). These AMGs can significantly influence global-scale microbial metabolism and are thus often searched for eagerly in newly discovered phage genomes. They can easily be misidentified, however, either because of the presence of a host region on a contig otherwise viral, or because of an error in the functional prediction of a new phage gene. Both types of errors are relatively frequent: host genes can easily be misconstrued as viral, especially when residing on the edge of a contig, while function prediction for genes only distantly related to references is notoriously difficult (see above). Ultimately, this type of error means that AMG identification on metagenome-assembled phage genomes cannot be based on an entirely automated process and requires systematic manual inspection of both the viral origin and predicted function of the candidate AMG.

2.7 CONCLUSION

In less than a decade, metagenomics has established itself as the approach of choice for phage diversity exploration. Tools for phage sequence identification and analysis are quickly emerging, while at the same time the community has started to established standards and guidelines (Roux et al. 2019). These tools, approaches, and best practices remain highly dynamic, however, especially with the fast development and adoption of new technologies such as long-reads sequencing. In addition, many challenges remain before the entire phage metagenomic pipeline is easily accessible to all researchers. Some of these challenges are linked to our limited knowledge of phage diversity: despite the impressive number of sequences reported in databases such as IMG/VR (Paez-Espino et al. 2018), entire parts of the virosphere are still vastly under-sampled and thus under-detected in metagenomes, including in particular RNA phages (Starr et al. 2019; Callanan et al. 2020). Other challenges are linked to inherent limitations of bioinformatics: in particular, in silico functional annotation and host prediction are dependent on the quality of the underlying databases, and will thus require a sustained effort to experimentally characterize and validate the function of novel phage genes as well as experimentally link uncultivated phages and hosts, before seeing a significant increase in efficiency. Finally, all these tools and resources must be made available not only to bioinformatics and metagenome specialists but also to researchers who may be less familiar with these techniques through user-friendly and highly documented interfaces. This work has been started through initiatives such as iVirus (ivirus.us (Bolduc et al. 2017)) and VERVENet (Kindler et al. 2016) and must be sustained in order to share the benefits provided by metagenome-based phage studies as widely as possible.

REFERENCES

Aggarwala, Varun, Guanxiang Liang, and Frederic D. Bushman. 2017. "Viral communities of the human gut: Metagenomic analysis of composition and dynamics." *Mobile DNA*, 1–10. doi: 10.1186/s13100-017-0095-y.

Ahlgren, Nathan A., Jie Ren, Yang Young Lu, Jed A. Fuhrman, and Fengzhu Sun. 2016. "Alignment-free d_2^* oligonucleotide frequency dissimilarity measure improves prediction of hosts from metagenomically-derived viral sequences." *Nucleic Acids Research* 45 (1): 39–53. doi: 10.1093/nar/gkw1002.

Akhter, Sajia, Ramy K. Aziz, and Robert A. Edwards. 2012. "PhiSpy: A novel algorithm for finding prophages in bacterial genomes that combines similarity- and composition-based strategies." *Nucleic Acids Research* 40 (16): 1–13.

Al-Shayeb, Basem, Rohan Sachdeva, Lin Xing Chen, Fred Ward, Patrick Munk, Audra Devoto, Cindy J. Castelle, et al. 2020. "Clades of huge phages from across Earth's ecosystems." *Nature* 578 (7795): 425–31. doi: 10.1038/s41586-020-2007-4.

Amgarten, Deyvid, Lucas P. P. Braga, Aline M. da Silva, and João C. Setubal. 2018. "MARVEL, a tool for prediction of bacteriophage sequences in metagenomic bins." *Frontiers in Genetics* 9 (August): 1–8. doi: 10.3389/fgene.2018.00304.

Anantharaman, Karthik, Melissa B. Duhaime, John A. Breier, Kathleen Wendt, Brandy M. Toner, and Gregory J. Dick. 2014. "Sulfur oxidation genes in diverse deep-sea viruses." *Science* 344 (6185): 757–60. doi: 10.3354/meps145269.

Antonov, Ivan, and Mark Borodovsky. 2010. "GeneTack: Frameshift identification in protein-coding sequences by the Viterbi algorithm." *Journal of Bioinformatics and Computational Biology* 8 (3): 535–51. doi: 10.1142/S0219720010004847.

Antonov, Ivan, Arthur Coakley, John F. Atkins, Pavel V. Baranov, and Mark Borodovsky. 2013. "Identification of the nature of reading frame transitions observed in prokaryotic genomes." *Nucleic Acids Research* 41 (13): 6514–30. doi: 10.1093/nar/gkt274.

Arndt, David, Jason R. Grant, Ana Marcu, Tanvir Sajed, Allison Pon, Yongjie Liang, and David S. Wishart. 2016. "PHASTER: A better, faster version of the PHAST phage search tool." *Nucleic Acids Research* 44 (May): 1–6. doi: 10.1093/nar/gkw387.

Baranov, Pavel V., Olivier Fayet, Roger W. Hendrix, and John F. Atkins. 2006. "Recoding in bacteriophages and bacterial IS elements." *Trends in Genetics* 22 (3): 174–81. doi: 10.1016/j.tig.2006.01.005.

Besemer, John, and Mark Borodovsky. 1999. "Heuristic approach to deriving models for gene finding." *Nucleic Acids Research* 27 (19): 3911–20. doi: 10.1093/nar/27.19.3911.

Bolduc, Benjamin, Ken Youens-Clark, Simon Roux, Bonnie L. Hurwitz, and Matthew B. Sullivan. 2017. "IVirus: Facilitating new insights in viral ecology with software and community data sets imbedded in a cyberinfrastructure." *The ISME Journal* 11 (1): 7–14. doi: 10.1038/ismej.2016.89.

Bowers, Robert M., Nikos C. Kyrpides, Ramunas Stepanauskas, Miranda Harmon-Smith, Devin Doud, T. B. K. Reddy, Frederik Schulz, et al. 2017. "Minimum Information about a Single Amplified Genome (MISAG) and a Metagenome-Assembled Genome (MIMAG) of bacteria and archaea." *Nature Biotechnology* 35 (8): 725–31. doi: 10.1038/nbt.3893.

Breitbart, Mya, Chelsea Bonnain, Kema Malki, and Natalie A. Sawaya. 2018. "Phage puppet masters of the marine microbial realm." *Nature Microbiology* 3 (July): 754–66. doi: 10.1038/s41564-018-0166-y.

Brum, Jennifer R., and Matthew B. Sullivan. 2015. "Rising to the challenge: Accelerated pace of discovery transforms marine virology." *Nature Reviews Microbiology* 13 (February): 1–13. doi: 10.1038/nrmicro3404.

Callanan, Julie, Stephen R. Stockdale, Andrey Shkoporov, Lorraine A. Draper, R. Paul Ross, and Colin Hill. 2018. "RNA phage biology in a metagenomic era." *Viruses* 10 (7): 1–17. doi: 10.3390/v10070386.

Callanan, Julie, Stephen R. Stockdale, Andrey. Shkoporov, Lorraine A. Draper, R. Paul Ross, and Colin Hill. 2020. "Expansion of known ssRNA phage genomes: From tens to over a thousand." *Science Advances* 6 (6). doi: 10.1126/sciadv.aay5981.

Delcher, Arthur L., Kirsten A. Bratke, Edwin C. Powers, and Steven L. Salzberg. 2007. "Identifying bacterial genes and endosymbiont DNA with glimmer." *Bioinformatics* 23 (6): 673–79. doi: 10.1093/bioinformatics/btm009.

Devoto, Audra E., Joanne M. Santini, Matthew R. Olm, Karthik Anantharaman, Patrick Munk, Jenny Tung, Elizabeth A. Archie, et al. 2019. "Megaphages infect Prevotella and variants are widespread in gut microbiomes." *Nature Microbiology* 4 (4): 693–700. doi: 10.1038/s41564-018-0338-9.

Dion, Moïra B., Frank Oechslin, and Sylvain Moineau. 2020. "Phage diversity, genomics and phylogeny." *Nature Reviews Microbiology* 18 (3): 125–38. doi: 10.1038/s41579-019-0311-5.

Edwards, Robert A., and Forest Rohwer. 2005. "Viral metagenomics." *Nature Reviews Microbiology* 3 (6): 504–10. doi: 10.1038/nrmicro1163.

Edwards, Robert A., Katelyn McNair, Karoline Faust, Jeroen Raes, and Bas E. Dutilh. 2016. "Computational approaches to predict bacteriophage-host relationships." *FEMS Microbiology Reviews* 40 (2): 258–72. doi: 10.1093/femsre/fuv048.

Emerson, Joanne B., Simon Roux, Jennifer R. Brum, Benjamin Bolduc, Ben J. Woodcroft, Ho Bin Jang, Caitlin M. Singleton, et al. 2018. "Host-linked soil viral ecology along a permafrost thaw gradient." *Nature Microbiology* 3 (8): 870–80. doi: 10.1038/s41564-018-0190-y.

Fancello, L., D. Raoult, and C. Desnues. 2012. "Computational tools for viral metagenomics and their application in clinical research." *Virology* 434 (2): 162–74. doi: 10.1016/j.virol.2012.09.025.

Finn, Robert D., Penelope Coggill, Ruth Y. Eberhardt, Sean R. Eddy, Jaina Mistry, Alex L. Mitchell, Simon C. Potter, et al. 2016. "The Pfam protein families database: Towards a more sustainable future." *Nucleic Acids Research* 44 (D1): D279–85. doi: 10.1093/nar/gkv1344.

Galiez, Clovis, Matthias Siebert, François Enault, Jonathan Vincent, and Johannes Söding. 2017. "WIsH: Who is the host? Predicting prokaryotic hosts from metagenomic phage contigs." *Bioinformatics* 33 (19): 3113–14. doi: 10.1093/bioinformatics/btx383.

Grazziotin, Ana Laura, Eugene V. Koonin, and David M. Kristensen. 2017. "Prokaryotic Virus Orthologous Groups (PVOGs): A resource for comparative genomics and protein family annotation." *Nucleic Acids Research* 45 (D1): D491–98. doi: 10.1093/nar/gkw975.

Gregory, Ann C., Ahmed A. Zayed, Nádia Conceição-Neto, Ben Temperton, Ben Bolduc, Adriana Alberti, Mathieu Ardyna, et al. 2019. "Marine DNA viral macro- and microdiversity from pole to pole." *Cell* 177 (5): 1109–23.e14. doi: 10.1016/j.cell.2019.03.040.

Hayes, Stephen, Jennifer Mahony, Arjen Nauta, and Douwe van Sinderen. 2017. "Metagenomic approaches to assess bacteriophages in various environmental niches." *Viruses* 9 (6): 127. doi: 10.3390/v9060127.

Hyatt, Doug, Gwo-liang Chen, Philip F. Locascio, Miriam L. Land, Frank W. Larimer, and Loren J. Hauser. 2010. "Prodigal : Prokaryotic gene recognition and translation initiation site identification." *BMC Bioinformatics* 11: 119.

Hyatt, Doug, Philip F. Locascio, Loren J. Hauser, and Edward C. Uberbacher. 2012. "Gene and translation initiation site prediction in metagenomic sequences." *Bioinformatics* 28 (17): 2223–30. doi: 10.1093/bioinformatics/bts429.

Ignacio-espinoza, J. Cesar, Nathan A. Ahlgren, and Jed A. Fuhrman. 2020. "Long-term stability and red queen-like strain dynamics in marine viruses." *Nature Microbiology*. doi: 10.1038/s41564-019-0628-x.

Ivanova, Natalia N., Patrick Schwientek, H. James Tripp, Christian Rinke, Amrita Pati, Marcel Huntemann, Axel Visel, Tanja Woyke, Nikos C. Kyrpides, and Edward M. Rubin. 2014. "Stop codon reassignments in the wild." *Science* 344 (6186): 909–13. doi: 10.1126/science.1250691.

Kanehisa, Minoru, Yoko Sato, Masayuki Kawashima, Miho Furumichi, and Mao Tanabe. 2016. "KEGG as a reference resource for gene and protein annotation." *Nucleic Acids Research* 44 (D1): D457–62. doi: 10.1093/nar/gkv1070.

Kelley, D. R., B. Liu, A. L. Delcher, and M. Pop. 2012. "Gene prediction with glimmer for metagenomic sequences augmented by classification and clustering." *Nucleic Acids Research* 40 (1): e9. http://nar.oxfordjournals.org/content/40/1/e9.short.

Kieft, Kristopher, Zhichao Zhou, and Karthik Anantharaman. 2020. "VIBRANT: Automated recovery, annotation and curation of microbial viruses, and evaluation of viral community function from genomic sequences." *Microbiome* 8 (1): 90. doi: 10.1186/s40168-020-00867-0.

Kindler, Lori, Alexei Stoliartchouk, Celina Gomez, James Thornton, Lenny Teytelman, and Bonnie L. Hurwitz. 2016. "VERVENet: The viral ecology research network." *PeerJ Preprints*. doi: 10.1890/15-0688.

Li, Dinghua, Ruibang Luo, Chi Man Liu, Chi Ming Leung, Hing Fung Ting, Kunihiko Sadakane, Hiroshi Yamashita, and Tak Wah Lam. 2016. "MEGAHIT v1.0: A fast and scalable metagenome assembler driven by advanced methodologies and community practices." *Methods* 102: 3–11. doi: 10.1016/j.ymeth.2016.02.020.

Lima-Mendez, Gipsi, Jacques Van Helden, Ariane Toussaint, and Raphaël Leplae. 2008. "Prophinder: A computational tool for prophage prediction in prokaryotic genomes." *Bioinformatics* 24 (6): 863–65. doi: 10.1093/bioinformatics/btn043.

Lomsadze, Alexandre, Karl Gemayel, Shiyuyun Tang, and Mark Borodovsky. 2018. "Modeling leaderless transcription and atypical genes results in more accurate gene prediction in prokaryotes." *Genome Research* 28 (7): 1079–89. doi: 10.1101/gr.230615.117.

Marchler-Bauer, Aron, Myra K. Derbyshire, Noreen R. Gonzales, Shennan Lu, Farideh Chitsaz, Lewis Y. Geer, Renata C. Geer, et al. 2015. "CDD: NCBI's conserved domain database." *Nucleic Acids Research* 43 (D1): D222–26. doi: 10.1093/nar/gku122.

Nayfach, Stephen, Antonio Pedro Camargo, Frederik Schulz, Emiley Eloe-fadrosh, Simon Roux, and Nikos Kyrpides. 2020. "CheckV: Assessing the quality of metagenome-assembled viral genomes." *Nature Biotechnology*, 1546–1696. doi: 10.1038/s41587-020-00774-7.

Noguchi, Hideki, Takeaki Taniguchi, and Takehiko Itoh. 2008. "MetaGeneAnnotator: Detecting species-specific patterns of ribosomal binding site for precise gene prediction in anonymous prokaryotic and phage genomes." *DNA Research* 15 (6): 387–96. doi: 10.1093/dnares/dsn027.

Nurk, Sergey, Dmitry Meleshko, Anton Korobeynikov, and Pavel A. Pevzner. 2017. "MetaSPAdes: A new versatile metagenomic assembler." *Genome Research* 5: 824–34. doi: 10.1101/gr.213959.116.

Páez-Espino, David, I-Min A. Chen, Krishna Palaniappan, Anna Ratner, Ken Chu, Ernest Szeto, Manoj Pillay, et al. 2016. "IMG/VR: A database of cultured and uncultured DNA viruses and retroviruses." *Nucleic Acids Research* 45 (October 2016): D457–65. doi: 10.1093/nar/gkw1030.

Paez-Espino, David, Simon Roux, I-Min A. Chen, Krishna Palaniappan, Anna Ratner, Ken Chu, Marcel Huntemann, et al. 2018. "IMG/VR v.2.0: An integrated data management and analysis system for cultivated and environmental viral genomes." *Nucleic Acids Research* 47 (D1): D678–86. doi: 10.1093/nar/gky1127.

Peng, Yu, Henry C. M. Leung, S. M. Yiu, and Francis Y. L. Chin. 2012. "IDBA-UD: A de novo assembler for single-cell and metagenomic sequencing data with highly uneven depth." *Bioinformatics* 28(11): 1420–28.

Ponsero, Alise J., and Bonnie L. Hurwitz. 2019. "The promises and pitfalls of machine learning for detecting viruses in aquatic metagenomes." *Frontiers in Microbiology* 10 (April): 1–6. doi: 10.3389/fmicb.2019.00806.

Ren, Jie, Nathan A. Ahlgren, Yang Young Lu, Jed A. Fuhrman, and Fengzhu Sun. 2017. "VirFinder: A Novel k-Mer based tool for identifying viral sequences from assembled metagenomic data." *Microbiome* 5 (69): 1–20. doi: 10.1186/s40168-017-0283-5.

Ren, Jie, Kai Song, Chao Deng, Nathan A. Ahlgren, Jed A. Fuhrman, Yi Li, Xiaohui Xie, Ryan Poplin, and Fengzhu Sun. 2020. "Identifying viruses from metagenomic data using deep learning." *Quantitative Biology* 8 (1): 64–77. doi: 10.1007/s40484-019-0187-4.

Rho, Mina, Haixu Tang, and Yuzhen Ye. 2010. "FragGeneScan: Predicting genes in short and error-prone reads." *Nucleic Acids Research* 38 (20): 1–12. doi: 10.1093/nar/gkq747.

Roux, Simon, Mart Krupovic, Didier Debroas, Patrick Forterre, and François Enault. 2013. "Assessment of viral community functional potential from viral metagenomes may be hampered by contamination with cellular sequences." *Open Biology* 3: 130160.

Roux, Simon, Francois Enault, Bonnie L. Hurwitz, and Matthew B. Sullivan. 2015. "VirSorter: Mining viral signal from microbial genomic data." *PeerJ* 3 (May): e985. doi: 10.7717/peerj.985.

Roux, Simon, Jennifer R. Brum, Bas E. Dutilh, Shinichi Sunagawa, Melissa B. Duhaime, Alexander Loy, Bonnie T. Poulos, et al. 2016. "Ecogenomics and potential biogeochemical impacts of uncultivated globally abundant ocean viruses." *Nature* 537 (7622): 689–93. doi: 10.1038/nature19366.

Roux, Simon, Evelien M. Adriaenssens, Bas E. Dutilh, Eugene V. Koonin, Andrew M. Kropinski, Mart Krupovič, Jens H. Kuhn, et al. 2019. "Minimum Information about an Uncultivated Virus Genome (MIUViG)." *Nature Biotechnology* 37: 29–37.

Shkoporov, Andrey N., and Colin Hill. 2019. "Bacteriophages of the human gut : The "known unknown" of the microbiome." *Cell Host and Microbe* 25 (2): 195–209. doi: 10.1016/j.chom.2019.01.017.

Simmonds, Peter, Mike J. Adams, Mária Benkő, Mya Breitbart, J. Rodney Brister, Eric B. Carstens, Andrew J. Davison, et al. 2017. "Consensus statement: Virus taxonomy in the age of metagenomics." *Nature Reviews Microbiology* 15: 161–68. doi: 10.1038/nrmicro.2016.177.

Starr, Evan P., Erin E. Nuccio, Jennifer Pett-Ridge, Jillian F. Banfield, and Mary K. Firestone. 2019. "Metatranscriptomic reconstruction reveals RNA viruses with the potential to shape carbon cycling in soil." *Proceedings of the National Academy of Sciences of the United States of America* 116 (51): 25900–908. doi: 10.1073/pnas.1908291116.

Sullivan, Matthew B. 2015. "Viromes, not gene markers for studying DsDNA viral communities." *Journal of Virology* 89 (5): 2459–61. doi: 10.1128/JVI.03289-14.

Sutton, Thomas D. S., and Colin Hill. 2019. "Gut bacteriophage: Current understanding and challenges." *Frontiers in Endocrinology* 10 (November): 1–18. doi: 10.3389/fendo.2019.00784.

Trubl, Gareth, Paul Hyman, Simon Roux, and Stephen T. Abedon. 2020. "Coming-of-age characterization of soil viruses: A user's guide to virus isolation, detection within metagenomes, and viromics." *Soil Systems* 4 (2): 23. doi: 10.3390/soilsystems4020023.

Wang, Weili, Jie Ren, Kujin Tang, Emily Dart, Julio Cesar Ignacio-Espinoza, Jed A. Fuhrman, Jonathan Braun, Fengzhu Sun, and Nathan A. Ahlgren. 2019. "A network-based integrated framework for predicting virus-host interactions." *BioRxiv* 2 (2): 505768. doi: 10.1101/505768.

Warwick-Dugdale, Joanna, Natalie Solonenko, Karen Moore, Lauren Chittick, Ann C. Gregory, Michael J. Allen, Matthew B. Sullivan, and Ben Temperton. 2019. "Long-read viral metagenomics captures abundant and microdiverse viral populations and their niche-defining genomic Islands." *PeerJ* 2019 (4): 1–28. doi: 10.7717/peerj.6800.

Williamson, Kurt E., Jeffry J. Fuhrmann, K. Eric Wommack, and Mark Radosevich. 2017. "Viruses in soil ecosystems: An unknown quantity within an unexplored territory." *Annual Review of Virology* 4 (1): 201–19. doi: 10.11 46/annurev-virology-101416-041639.

Yutin, Natalya, Sean Benler, Sergei Shmakov, Yuri Wolf, Igor Tolstoy, Mike Rayko, Dmitry Antipov, Pavel Pevzner, and Eugene Koonin. 2020. "Unique genomic features of CrAss-like phages, the dominant component of the human gut virome." BioRxiv. doi: 10.1101/2020.07.20.212944.

Zhang, Ruoshi, Milot Mirdita, Eli Levy Karin, Clovis Norroy, Clovis Galiez, and Johannes Soeding. 2020. "SpacePHARER: Sensitive identification of phages from CRISPR spacers in prokaryotic hosts." BioRxiv. doi: 10.1101/2020.05.15.090266.

Zhu, Wenhan, Alexandre Lomsadze, and Mark Borodovsky. 2010. "Ab initio gene identification in metagenomic sequences." *Nucleic Acids Research* 38 (12): 1–15. doi: 10.1093/nar/gkq275.

CHAPTER 3

Direct RNA Sequencing for Complete Viral Genomes

Sebastian Krautwurst

Friedrich Schiller University Jena

Ronald Dijkman and Volker Thiel

University of Bern

Andi Krumbholz

Christian-Albrecht University of Kiel

Manja Marz

Friedrich Schiller University Jena

CONTENTS

3.1 ADVANTAGES AND DISADVANTAGES FOR VIRUSES

Although the first complete genome to be sequenced was the viral RNA-genome Escherichia virus MS2 (Fiers et al., 1976), most sequencing approaches of the last years have been developed for prokaryotes or eukaryotes. Arguably the greatest leap in sequencing technology was the advent of high-throughput sequencing (HTS; also called the next-generation sequencing, NGS) (Goodwin et al., 2016).

Short-Read HTS technologies by Roche 454, IonTorrent, and Solexa/ Illumina have enabled sequencing of millions of short fragments in a highly parallel manner (Rothberg and Leamon, 2008; Merriman et al., 2012; Quail et al., 2008). Modern Illumina machines achieve excellent accuracy that enables robust detection of single nucleotide polymorphisms (SNPs), and thus facilitates the characterization of viral mutations and haplotypes. This technology provides very high throughput, reducing the costs of sequencing a complete human genome below $1,000. Main drawbacks include a limited read length (maximum of 400 nt) which can be insufficient to resolving complex genomic structures, e.g., long repeats. The methods can only sequence DNA, and thus must fall back on cDNA sequencing for sequencing of RNA viruses or for transcriptomic analyses. Downstream analyses relying on read quantification can suffer from bias introduced by read amplification steps. This is especially true for viruses, due to an inhomogeneous amplification of different viruses (Sujayanont et al., 2014).

Single-cell sequencing is a recent advancement that exploits the capabilities of NGS. The technology by 10x Genomics enables isolating single cells in oil droplets in which the cell is lysed, and all fragments are tagged and prepared for Illumina sequencing (Zhang et al., 2019). Single-cell sequencing approaches have helped to assess virus genome diversity and virus–host interactions in the presence of intercell heterogeneity during virus infections (Ciuffi et al., 2016), e.g. for flaviviruses (Zanini et al., 2018).

Recent years have brought the advent of widely available long-read sequencing technologies, which can determine the nucleotide sequence of longer fragments. The first widely available method was Pacific Biosciences' single-molecule real-time (SMRT) sequencing (Eid et al., 2009). Current SMRT flowcells can yield up to 500 Gb of uncorrected sequence throughput with typical read lengths of around 15 kb, or around 65 Gb when most fragments are longer than 50 kb. Uncorrected reads have an error rate of 15%, which can be reduced to 0.1% by sequencing fragments multiple times in a circular fashion, albeit at the cost of reducing throughput by 90%. The method is restricted to DNA sequencing, but libraries comprised of complementary DNA (cDNA) can be utilized to infer RNA sequences. SMRT sequencing has the ability to detect DNA modifications. The long-read length makes this technology suitable to sequence full viral sequences. It can be combined with single-cell sequencing methods, e.g., to sequence influenza virions infecting single cells (Russell et al., 2019).

Oxford Nanopore Technologies (ONT) has developed a nanopore sequencing method that works by translocating nucleotide strands through a multi-protein pore complex which forms a channel through a membrane (see Figure 3.1) and measuring the shifts of the electrical current caused by the passing nucleotides. The complex was derived from a natural bacterial "blueprint" and further developed and improved to best meet requirements of the technology. The translocation is achieved by applying a voltage across the membrane in which the nanopore is placed. A motor protein which is attached to every strand during library preparation is utilized to direct and smoothen the traversal through the pore. At the core of the complex, a sensor measures the electrical current flowing through the pore. This current is influenced by the nucleotides passing through the core of the complex. Nanopores are arranged on flowcells in a large array to allow highly parallel sequencing, yielding up to 25 Gb from up to 1,600 pores by sequencing on the MinION device. Larger-scale devices and flowcells are available (PromethION). Barcoding kits are available, allowing for loading of up to

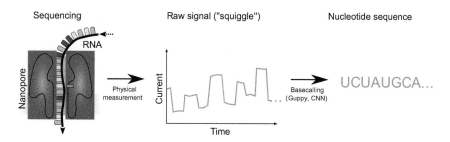

FIGURE 3.1 Nanopore sequencing workflow. Nucleotide strands are translocated through a multi-protein pore complex which forms a channel through a membrane. Inside the complex sits a sensor that measures the electrical current at the center of the channel. The current is influenced by the nucleotides passing through. The raw signal measurements (called "squiggle") need to wbe converted to the nucleotide sequence ("basecalling") with suitable software, e.g., **Guppy**, which uses convolutional neural networks (CNN).

96 samples on a single flowcell, thus reducing per-sample sequencing costs by a large factor if lower read numbers are sufficient for the application (Jain et al., 2016).

The raw current information generated by the sequencer (called a "squiggle") needs to be translated to the nucleotide sequence in a separate process termed "basecalling." State-of-the-art basecalling software (**Guppy** basecaller, available from the Oxford Nanopore Community) utilizes complex convolutional neural networks to achieve this task with high accuracy, which is computationally very taxing and best tackled with powerful graphics processing unit hardware (Wick et al., 2019). This disconnection between the raw data generation and basecalling steps has the unique advantage that any software improvements for the computational latter part can retroactively be applied to all raw data generated in previous sequencing experiments. This allows accuracy improvements to carry over to all nanopore sequencing data from the past.

Nanopore sequencing technology has enabled direct RNA sequencing (DRS) for the first time. This avoids multiple drawbacks of previous sequencing methods arising from cDNA synthesis and amplification, such as reverse transcription bias, polymerase chain reaction (PCR) bias, or the generation of artificial RNA-RNA chimeras. These chimeric sequences cannot be easily distinguished from naturally occurring recombinant architectures such as RNAs resulting from discontinuous replication in viruses. Also, not including any amplification step prior to sequencing retains all RNA modifications in the input sample, whereas this information is lost in all other approaches. As such, nanopore DRS is uniquely viable to find the complete spectrum of recombination variants and haplotypes of complex RNA populations such as viral quasispecies. Sequencing yield of the direct RNA protocol for one flowcell on the MinION device can be upwards of 5 Gb but is highly dependent on the input sample concentration. At least 500ng of input RNA is recommended, of which all sequences have to be polyadenylated, otherwise the adapters needed for sequencing cannot be attached. For RNA without polyA, it is required to employ a polyA-tailing kit or to design and ligate custom adapter sequences to 3′ ends. Direct RNA sequencing is of utmost importance in the field of virology. RNA viruses can directly be sequenced without amplification, which opens the door to describe haplotypes (Li et al., 2016; Harel et al., 2019), the entire quasispecies (Li et al., 2016), structural variants (Viehweger et al., 2019; Depledge et al., 2019), but also to assess quantity measures (Viehweger et al., 2019), and RNA modifications (Stoiber et al., 2016; Pratanwanich et al., 2020).

The basecalling process is error-prone even for current state-of-the-art basecalling programs, as the measured raw signal is subject to high variance, while the induced changes due to nucleotides present in the pore are only slight. State-of-the-art basecalling software uses large-scale convolutional neural networks which work directly on the raw signal to predict the most probable underlying nucleotide sequence. The neural networks are trained on a large collection of sequence data comprised of many different organisms, but results can still vary a lot depending on whether the analyzed RNA is similar to sequences the network model was trained on or not (Rang et al., 2018; Viehweger et al., 2019).

For DNA, modal basecalling error rates are typically between 3% and 5% for single raw reads (Rang et al., 2018; ONT, 2021). In RNA basecalling, single read error rates vary from 8% to 15%, depending on source organisms and presumably the presence of RNA modifications (see Table 3.1) (Rang et al., 2018; Viehweger et al., 2019; Parker et al., 2020; ONT, 2021). When sufficient reads of the same sequence are available, these can be combined to average out the random errors that occur during basecalling, reducing error rates to 0.1% and below in the resulting consensus sequence (Wick et al., 2019; ONT, 2021). These values represent the state of the technology at the time of writing of this book but are expected to be improved continuously in the future as new algorithms are developed. A new approach based on Connectionist Temporal Classification (CTC) is bound to improve upon current basecalling methods (bonito basecaller, ONT) (ONT, 2020a). Also, a new version of nanopores with increased sensitivity to current changes is close to a full release (R10 pore model). Prototypes of

TABLE 3.1 Nanopore DRS Error Statistics

Sample	Mapped to	%subst.	%insert.	%delet.	%total
HCoV[a]	HCoV-229E (NC002645.1)	2.30	2.00	5.01	9.32
	Homo sapiens (GRCh38)	2.46	2.41	5.06	9.94
Vero[b]	SARS-CoV-2 (NC045512.2)	3.19	2.77	5.27	11.23
HNC[c]	Homo sapiens (GRCh38)	4.26	2.84	6.79	13.88

[a] Total RNA of HuH7 cells infected with Human coronavirus 229E.
[b] Total RNA of vero cells (generated from the kidney of an African green monkey) infected with severe acute respiratory syndrome coronavirus type 2 (SARS-CoV-2).
[c] Total RNA from human head and neck cancer tissue cells. Each sample was sequenced with the direct RNA protocol on the Oxford Nanopore Technologies MinION. Raw data was basecalled with Guppy (v4.0.11) and mapped to the respective reference genomes with minimap2. Note that all error rates report raw differences to the reference genome and thus include actual genetic variation.

this model already surpass the current R9 nanopores in sequencing accuracy (ONT, 2020b).

3.2 VIRUS ASSEMBLY OF THE HUMAN CORONAVIRUS 229E

With long-read sequencing, it is possible to generate reads which span complete viral genomes. As an example, employing nanopore DRS on a sample of total RNA extracted from Huh7 cells infected with Human Coronavirus 229E (HCoV-229E) yielded multiple reads that represented nearly the complete genome with a size of about 27,300 nt.

Coronaviruses are enveloped single-stranded positive-sense RNA viruses that infect various mammalian and avian hosts (Vijay and Perlman, 2016). Recent zoonoses from animals to humans, including the ongoing severe acute respiratory syndrome coronavirus type 2 (SARS-CoV-2) pandemic, highlight the medical importance of understanding the replication mechanics of these viruses (V'kovski et al., 2020). Coronaviruses possess genomes larger than most other RNA virus genomes. While long reads are very well suited for any assembly task, they are especially useful for the study of very complex genomes that show frequent recombination, subgenomic isoforms or long repeat regions, etc. Many virus genomes, particularly of RNA viruses, involve complicated architectures or include such patterns during their replication mechanisms.

The replication of coronaviruses includes a process, called discontinuous extension of minus strands, to produce a nested set of 5'- and 3'-coterminal subgenomic RNAs (sg RNAs). These carry a common 5' leader sequence that is identical to the 5' end of the viral genome. The sg RNAs contain a different number of open reading frames (ORFs) that encode the viral structural proteins and several accessory proteins. With very few exceptions, only the 5'-located ORF (which is absent from the next smaller sg RNA) is translated into protein (see Figure 3.2) (Schreiber et al., 1989).

These complex architectural details for viral sg RNAs produced in cells infected with HCoV-229E can be assessed using DRS. Complete viral mRNAs, including the full coronavirus genome, can be sequenced in single contiguous reads. Sequence analysis of thousands of full-length sg RNAs allows examination of the leader-body junction sites of the major viral mRNAs. In addition, this approach provides insight into the diversity of additional HCoV-229E sg RNAs, probably including defective interfering RNA (DI-RNAs).

To this end, the reads need to be aligned to the reference genome via a mapping program. Most commonly used for nanopore reads is

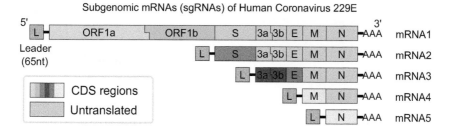

Subgenomic mRNAs (sgRNAs) of Human Coronavirus 229E

FIGURE 3.2   Overview of mRNAs produced in cells infected with Human coronavirus 229E. A mechanism termed discontinuous replication of negative strands gives rise to several subgenomic mRNAs which contain a copy of the 65 nt leader sequence at the 5′ end (Thiel et al., 2003). Each mRNA is used to produce only the proteins corresponding to the colored 5′-proximal coding sequence (CDS).

minimap2, which is a versatile sequence alignment program that supports DNA and RNA mapping. It also supports spliced alignments. The main advantage of **minimap2** is that it is considerably faster than other programs with no or minimal loss in mapping accuracy or sensitivity. This is achieved by using fast index structures that contain so-called minimizers instead of exact k-mers from the reference. Minimizers are a lossy representation of a group of similar k-mers, and thus a smaller number of them represent the whole reference. During alignment, this speeds up the search for suitable seed regions considerably. Alignments are then maximally extended in both directions as usual.

Mapping the reads to the reference reveals the coverage of the genome achieved by sequencing. Plots depicting the coverage can be produced from the alignment file (.sam or.bam). These can reveal interesting facts about the sample composition, e.g., fragment or isoform abundances. For three DRS runs of the murine coronavirus A59, the plot shows a large coverage bias toward both ends (see Figure 3.3). The high coverage of the 3′-end reflects the higher abundance of mRNAs produced from the 3′-terminal genome regions and is a result of the discontinuous transcription mechanism employed by coronaviruses and several other nidoviruses (Sawicki and Sawicki, 1995). The 3′-coverage is further increased by the directional sequencing that starts from the sg RNA 3′-terminal polyA tail. Also, the observed coverage bias for the very 5′-end results from the coronavirus-specific transcription mechanism because all viral mRNAs are equipped with the 65nt 5′-leader sequence derived from the 5′-end of the genome. However, a portion of the reads does not show a split alignment including the leader sequence, possibly due to failed completion of

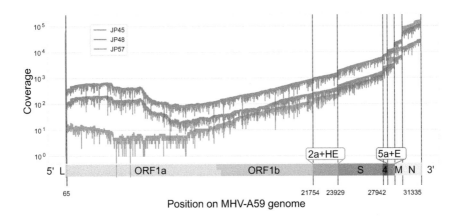

FIGURE 3.3 Genome coverage plot for three murine coronavirus (MHV) direct RNA nanopore sequencing datasets. The high coverage toward the 3'-end arises from the higher abundance of subgenomic RNAs (sg RNAs) produced from the 3'-terminal genome regions due to the discontinuous transcription mechanism of coronaviruses (Sawicki and Sawicki, 1995). Distinct "steps" in the coverage correspond to the borders of known sg RNAs (vertical dotted lines).

the mechanism or mapping errors. The remainder of the high 5'-coverage bias likely reflects the presence of high numbers of DI-RNAs in which 5'- and 3'-proximal genomic sequences were fused, probably resulting from illegitimate recombination events as shown previously for other coronaviruses.

It is also possible to generate a consensus sequence from the sequencing reads. After the alignment to a reference, the relative position of all mapped reads is confidently determined. Consensus programs such as **Medaka** will select a nucleotide for every position in the reference based on the information from all spanning reads. This process corrects all singular random sequencing errors from the reads. Additionally, any systematic changes in the input read data, such as substitutions (including SNPs), insertions, or deletions, will be included in the consensus sequence. This information is vital for studying evolutionary changes, e.g., during virus outbreaks.

3.3 LONG READS ENABLE DISCOVERY OF LONG-RANGE INTERACTIONS AND GENOME-WIDE COMPENSATORY MUTATIONS

Another inherent aspect of long-reads is the possibility to find co-occurring mutations without inferring them from short reads. Especially for RNA viruses, RNA secondary structures play an essential role in various stages of the viral life cycle. For example, the replication of Dengue viruses depends on a long-range RNA–RNA interaction (LRI) between the 5′ UTR and 3′ UTR of the viral genome (Hahn et al., 1987; Alvarez et al., 2005). With conventional NGS experiments (RNA-Seq, Illumina), it was impossible to create reads that cover both UTRs, since the genome of Dengue virus is ~10,000 nt long. With long-read technologies, on the other hand, the possibility to obtain reads covering such long genomes is given and LRIs can be potentially found in the reads directly. For this, reads are scanned for co-occurring mutations that do not disrupt a hypothetical underlying RNA secondary structure. Such a pair of mutations is also called a compensatory mutation. For each compensatory mutation, all covering reads can be aligned to create a multiple sequence alignment (MSA). Next, the flanking regions of the compensatory mutation are analyzed whether an RNA–RNA interaction is possible, usually based on thermodynamic energies and individual base-pairing probabilities. Due to the high mutation rate of viruses, compensatory mutations might be detectable in patient samples. Therefore, long-read technologies enable the direct analyses of LRIs in viral read data and thus may help unravel the mysteries of viral replication and viral regulation during host infection.

3.4 SEQUENCING FULL RNA VIRAL TRANSCRIPTS

Long-read sequencing is uniquely capable of accurately capturing complete DNA and RNA fragments in single contiguous sequences. This allows quantification of different transcripts in complex genomic architectures such as differently spliced mRNAs. With short-read sequencing, it is generally impossible to assign reads to these transcript variants unambiguously.

The discontinuous transcription mechanism employed by coronaviruses gives rise to several subgenomic RNAs that share sequence content 2. The aligned reads distribution reveals clusters for all known canonical mRNAs which closely fit the expected molecule lengths, showing that full-length mRNAs can be sequenced (see Figure 3.4).

Apart from known leader-body junctions present in canonical sg RNAs, a high number of other recombination sites can be observed which have not been described before (see Figure 3.5). A recombination site can be defined as any site adjacent to a gap with a minimum length of 100 nt found by discontinuous mapping. This demonstrates how the recovery of full transcripts with nanopore sequencing can give insights into complex

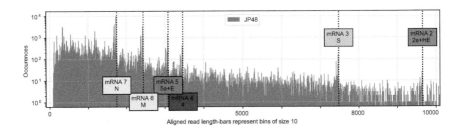

FIGURE 3.4   Full-length viral transcripts: Shown is the distribution of aligned read lengths for a direct RNA sequencing dataset of murine coronavirus (MHV) A59 strain. Total RNA of infected cells was sequenced with DRS, and reads were mapped to the MHV-A59 strain reference genome. The sample shows clusters that correspond to the lengths of known mRNAs (vertical dotted lines). The doubled peaks arise from transcripts that are lacking the leader sequence, either due to a disruption during the discontinuous replication mechanism or due to mapping errors caused by sequencing errors. Note the logarithmic scale of the *y*-axis.

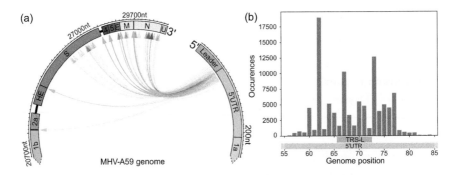

FIGURE 3.5   Leader-body junctions of murine coronavirus (MHV) sg RNAs. (a) The green arrows show occurrences of leader-body junctions corresponding to canonical sg RNAs, whereas red arrows show unknown junctions, which likely stem from DI-RNAs present in the sample. (b) Distribution of junction start sites. A large fraction of junctions start outside of the leader TRS region.

structural properties and mechanics, e.g., the discontinuous replication of coronaviruses. For samples of murine coronavirus (MHV) strain A59 infecting human cells, diverse noncanonical leader-body junctions were found, with some displaying high read support. This supports the hypothesis that the discontinuous replication mechanism is highly error-prone, giving rise to numerous DI-RNAs (Pathak and Nagy, 2009).

3.5 MODIFICATIONS

Nanopore sequencing preserves nucleotide modifications during sequencing, as it does not require amplification steps during library preparation. This is also true for the DRS protocol, which for the first time allows sequencing and detection of RNA modifications from the RNA strands as they appear in the sample. Modified nucleotides (e.g., methylated bases, pseudouridine) cause a different electric current shift than canonical nucleotides during pore traversal (Garalde et al., 2018). Thus, these bases generally complicate basecalling, as the models used for basecalling are trained on unmodified sequences. Basecalling models have recently been enhanced for the detection of certain methylation variants in the **Guppy** basecaller (5-Methylcytosine (5mC) and N^6-Methyladenosine (m6A), only on DNA).

Specialized tools for nucleotide modification detection on nanopore sequencing data have been developed. One of the first ones, **Tombo** by ONT (Stoiber et al., 2016), re-aligns the raw signal to the reference-corrected read sequence (called "re-squiggling"). For each base position, this yields the corresponding interval of the raw signal. The signal mean and standard deviation are then compared to known values of canonical bases in the same context to determine the methylation status. The final output consists of methylation rates for all positions on the reference genome/transcriptome, aggregated from all overlapping reads (see Figure 3.6). As raw nanopore sequencing data is subject to considerable variance, Tombo often struggles with overestimating the divergence from known canonical values, leading to false positives.

More recent programs aim to improve detection accuracy with more sophisticated approaches. For example, **xPore** (Pratanwanich et al., 2020) employs an improved method of re-squiggling, and then collects the distribution of current values for each reference position. After fitting a mixture of two Gaussian distributions to this data, the distribution closer to the known canonical mean of this nucleotide context is designated as unmodified, and the other one as modified. This allows estimating the fraction of modified reads at this position, as well as the modification probability of each read.

3.6 TRACKING VIRUS MUTATIONS DURING OUTBREAKS

Two important prerequisites for massively sequencing virus isolates during an outbreak situation, such as the SARS-CoV-2 pandemic that started in 2019, are low sequencing costs and ease of library generation. Recovering

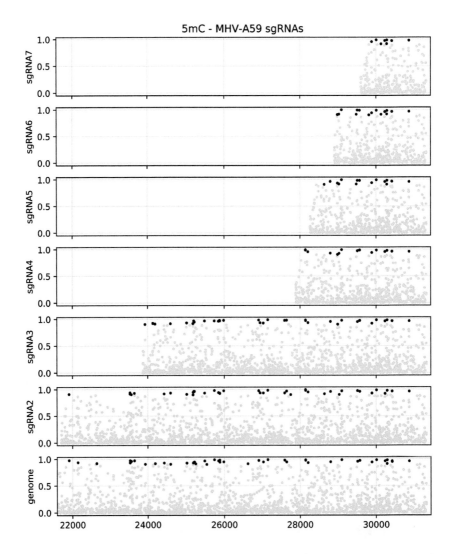

FIGURE 3.6   5mC methylation frequencies (determined by **Tombo** (Stoiber et al., 2016)) in MHV-A59 sg RNAs. Positions are marked in black if they were designated as methylated in 90% of overlapping reads. Some of the positions are consistently methylated across all sg RNAs.

viral RNA populations of coronavirus-infected cells by direct RNA nanopore sequencing of total cellular RNA is not ideal, as the copy number bias toward shorter sg RNAs leads to a highly uneven coverage of the genome 7. For accurate detection of SNPs, low-coverage regions need to be avoided by sequencing more of each sample, driving up costs. Also, sample multiplexing with barcoding is not natively supported for direct RNA.

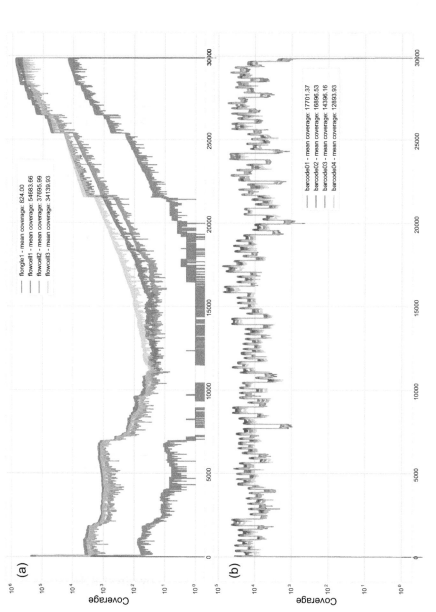

FIGURE 3.7 Sequencing coverage for SARS-CoV-2. (a) Direct RNA sequencing starts from the 3′ end of the coronaviruses. Not all reads are read to the 5′ end. (b) Cheaper amplicon sequencing for a homogeneous coverage, containing no information about haplo-types, quasispecies, compensatory mutations, and RNA modifications. Logarithmic scaling of y-axes (coverage).

To address these issues for the sequencing of SARS-CoV-2 samples, the ARTIC protocol was developed (Tyson et al., 2020). A set of specifically designed primers is used to create DNA fragments from overlapping genome regions by reverse transcription, which are then amplified by PCR. These amplicons display a very even coverage distribution (see Figure 3.7) and are thus ideal for assembly of the viral genome. Additionally, the amplicons being made up of DNA cause lower error rates and higher throughput with nanopore sequencing. High sequencing yields allow for up to 48 samples to be sequenced on one flowcell in parallel with barcoding, which brings total consumable costs per sample down to 80€. Compared to direct RNA sequencing, DNA amplicons do not allow detection of compensatory mutations and RNA modifications, and are not suited for quasispecies reconstruction.

ACKNOWLEDGMENTS

The authors thank Ruben Rose, Milena Žarković, and Jasmine Portmann for their help with genomic extraction and nanopore sequencing of the SARS-CoV-2 and MHV-A59 samples.

REFERENCES

Alvarez, D. E., M. F. Lodeiro, S. J. Luduena, L. I. Pietrasanta, and A. V. Gamarnik, 2005. Long-range RNA-RNA interactions circularize the dengue virus genome. *Journal of Virology* **79**: 6631–6643.

Ciuffi, A., S. Rato, and A. Telenti, 2016. Single-cell genomics for virology. *Viruses* **8**: 123.

Depledge, D. P., K. P. Srinivas, T. Sadaoka, D. Bready, Y. Mori, et al., 2019. Direct RNA sequencing on nanopore arrays redefines the transcriptional complexity of a viral pathogen. *Nature Communications* **10**: 1–13.

Eid, J., A. Fehr, J. Gray, K. Luong, J. Lyle, et al., 2009. Real-time DNA sequencing from single polymerase molecules. *Science* **323**: 133–138.

Fiers, W., R. Contreras, F. Duerinck, G. Haegeman, D. Iserentant, et al., 1976. Complete nucleotide sequence of bacteriophage MS2 RNA: Primary and secondary structure of the replicase gene. *Nature* **260**: 500–507.

Garalde, D. R., E. A. Snell, D. Jachimowicz, B. Sipos, J. H. Lloyd, et al., 2018. Highly parallel direct RNA sequencing on an array of nanopores. *Nature Methods* **15**: 201.

Goodwin, S., J. D. McPherson, and W. R. McCombie, 2016. Coming of age: Ten years of next-generation sequencing technologies. *Nature Reviews Genetics* **17**: 333.

Hahn, C. S., Y. S. Hahn, C. M. Rice, E. Lee, L. Dalgarno, et al., 1987. Conserved elements in the 3′ untranslated region of flavivirus RNAs and potential cyclization sequences. *Journal of Molecular Biology* **198**: 33–41.

Harel, N., M. Meir, U. Gophna, and A. Stern, 2019. Direct sequencing of RNA with MinION nanopore: Detecting mutations based on associations. *Nucleic Acids Research* **47**: e148.

Jain, M., H. E. Olsen, B. Paten, and M. Akeson, 2016. The Oxford Nanopore MinION: Delivery of nanopore sequencing to the genomics community. *Genome Biology* **17**: 239.

Li, C., K. R. Chng, E. J. H. Boey, A. H. Q. Ng, A. Wilm, et al., 2016. INC-Seq: Accurate single molecule reads using nanopore sequencing. *GigaScience* **5**: 34.

Merriman, B., Ion Torrent R&D Team, and J. M. Rothberg, 2012. Progress in ion torrent semiconductor chip based sequencing. *Electrophoresis* **33**: 3397–3417.

ONT, 2020a. New research algorithms yield accuracy gains for nanopore sequencing. https://nanoporetech.com/about-us/news/new-research-algorithms-yield-accuracy-gains-nanopore-sequencing, [Online; accessed 17-January-2021].

ONT, 2020b. R10.3: The newest nanopore for high accuracy nanopore sequencing – now available in store. https://nanoporetech.com/about-us/news/r103-newest-nanopore-high-accuracy-nanopore-sequencing-now-available-store, [Online; accessed 17-January-2021].

ONT, 2021. Accuracy. https://nanoporetech.com/accuracy, [Online; accessed 17-January-2021].

Parker, M. T., K. Knop, A. V. Sherwood, N. J. Schurch, K. Mackinnon, et al., 2020. Nanopore direct RNA sequencing maps the complexity of Arabidopsis mRNA processing and m6A modification. *Elife* **9**: e49658.

Pathak, K. B. and P. D. Nagy, 2009. Defective interfering RNAs: Foes of viruses and friends of virologists. *Viruses* **1**: 895–919.

Pratanwanich, P. N., F. Yao, Y. Chen, C. W. Koh, C. Hendra, et al., 2020. Detection of differential RNA modifications from direct RNA sequencing of human cell lines. *BioRxiv*.

Quail, M. A., I. Kozarewa, F. Smith, A. Scally, P. J. Stephens, et al., 2008. A large genome center's improvements to the Illumina sequencing system. *Nature Methods* **5**: 1005–1010.

Rang, F. J., W. P. Kloosterman, and J. de Ridder, 2018. From squiggle to basepair: Computational approaches for improving nanopore sequencing read accuracy. *Genome Biology* **19**: 90.

Rothberg, J. M. and J. H. Leamon, 2008. The development and impact of 454 sequencing. *Nature Biotechnology* **26**: 1117–1124.

Russell, A. B., E. Elshina, J. R. Kowalsky, A. J. Te Velthuis, and J. D. Bloom, 2019. Single-cell virus sequencing of influenza infections that trigger innate immunity. *Journal of Virology* **93**: e00500–19.

Sawicki, S. G. and D. L. Sawicki, 1995. Coronaviruses use discontinuous extension for synthesis of subgenome-length negative strands. In Corona-and Related Viruses, pp. 499–506, Springer: Boston, MA.

Schreiber, S. S., T. Kamahora, and M. M. Lai, 1989. Sequence analysis of the nucleocapsid protein gene of human coronavirus 229E. *Virology* **169**: 142–151.

Stoiber, M., J. Quick, R. Egan, J. E. Lee, S. Celniker, et al., 2016. De novo identification of DNA modifications enabled by genome-guided nanopore signal processing. BioRxiv, p. 094672.

Sujayanont, P., K. Chininmanu, B. Tassaneetrithep, N. Tangthawornchaikul, P. Malasit, et al., 2014. Comparison of phi29-based whole genome amplification and whole transcriptome amplification in dengue virus. *Journal of Virological Methods* **195**: 141–147.

Thiel, V., K. A. Ivanov, A. Putics, T. Hertzig, B. Schelle, et al., 2003. Mechanisms and enzymes involved in SARS coronavirus genome expression. *Journal of General Virology* **84**: 2305–2315.

Tyson, J. R., P. James, D. Stoddart, N. Sparks, A. Wickenhagen, et al., 2020. Improvements to the ARTIC multiplex PCR method for SARS-CoV-2 genome sequencing using nanopore. *BioRxiv.*

V'kovski, P., A. Kratzel, S. Steiner, H. Stalder, and V. Thiel, 2020. Coronavirus biology and replication: Implications for SARS-CoV-2. *Nature Reviews Microbiology* **19**: 1–16.

Viehweger, A., S. Krautwurst, K. Lamkiewicz, R. Madhugiri, J. Ziebuhr, et al., 2019. Direct RNA nanopore sequencing of full-length coronavirus genomes provides novel insights into structural variants and enables modification analysis. *Genome Research* **29**: 1545–1554.

Vijay, R. and S. Perlman, 2016. Middle east respiratory syndrome and severe acute respiratory syndrome. *Current Opinion in Virology* **16**: 70–76.

Wick, R. R., L. M. Judd, and K. E. Holt, 2019. Performance of neural network basecalling tools for Oxford Nanopore sequencing. *Genome Biology* **20**: 129.

Zanini, F., S.-Y. Pu, E. Bekerman, S. Einav, and S. R. Quake, 2018. Single-cell transcriptional dynamics of flavivirus infection. *Elife* **7**: e32942.

Zhang, X., T. Li, F. Liu, Y. Chen, J. Yao, et al., 2019. Comparative analysis of droplet-based ultra-high-throughput single-cell RNA-seq systems. *Molecular Cell* **73**: 130–142.

CHAPTER 4

Computational Methods for Viral Quasispecies Assembly

Kim Philipp Jablonski and Niko Beerenwinkel

ETH Zurich

SIB Swiss Institute of Bioinformatics

CONTENTS

4.1 INTRODUCTION

RNA viruses, such as HIV, HCV, Influenza virus, and SARS-CoV-2, are of major concern to public health. For example, SARS-CoV-2, the cause of COVID-19, has been declared a pandemic by the World Health Organization with 34 million confirmed cases and over a million deaths worldwide by September 2020 (Dong et al. 2020). RNA viruses occur in single-stranded and double-stranded forms and are characterized by large population sizes, small genomes featuring few proteins, high mutation rates due to the lack of proofreading capabilities of RNA polymerases, and short generation times (Duffy et al. 2008).

These factors lead to an extraordinary degree of genetic diversity, both on an inter- as well as intra-patient level. In particular, the set of non-identical but similar viral strains infecting a single host with varying abundances is called a viral quasispecies. It typically consists of one or a few haplotypes of high frequency and a large tail of low-frequency variants. The term "quasispecies" originates from a model of self-replicating macromolecules and refers to the mutation-selection equilibrium distribution of mutants (Eigen and Schuster 1977). The distribution of viral haplotypes influences a wide range of clinically relevant factors, such as escape from the host's immune response, vaccination success, drug resistance, and viral pathogenesis (Nowak et al. 1991; Gaschen et al. 2002; Johnson et al. 2008; Tsibris et al. 2009). Therefore, it is of great interest to assess, in a quantitative fashion, the viral quasispecies making up an infection.

Up until the last decade, Sanger sequencing was the most commonly used method to sequence cloned viruses obtained from plasma or infected cells of a host organism (Shendure et al. 2017; Beerenwinkel and Zagordi 2011). Due to it being labor- and time-intensive as well as yielding only low-coverage read sets, Sanger sequencing is only useful to determine consensus sequences and thus misses out on potentially important low-frequency haplotypes. By contrast, nowadays, next-generation sequencing (NGS) technologies provide a faster and cheaper alternative to generate high-coverage datasets (Beerenwinkel et al. 2012; Marz et al. 2014).

Given NGS sequencing data of viral RNA (or cDNA), genetic diversity estimation can be performed at three different scales. At the single-nucleotide variant (SNV) level, point mutations are detected, typically without incorporating linkage information between different genomic loci. At the local scale, consecutive genomic regions are considered that can be covered by the average sequencing read length such that local haplotypes can be estimated. This is not only useful when a constrained genomic region is of interest, but also for SNV calling by taking advantage of locally co-occurring mutations. At the global scale, a genomic region longer than the read length is of interest (e.g., whole viral genomes analyzed by short-read NGS technologies). This setting requires computational reconstruction of the global haplotypes, i.e., overlapping reads have to be merged to give rise to a full-length quasispecies reconstruction.

4.2 CHALLENGES OF GLOBAL HAPLOTYPE RECONSTRUCTION

While a high-quality global haplotype reconstruction would be most beneficial to understanding the previously mentioned clinically relevant factors, it comes with a number of challenges, which have not been solved so far. These issues already arise during sample preparation and sequencing (Posada-Cespedes et al. 2017; Beerenwinkel et al. 2012), but we will focus on computational aspects of the data analysis here (Beerenwinkel and Zagordi 2011; Töpfer and Beerenwinkel 2016; Seifert and Beerenwinkel 2016).

One of the most basic problems is that low-frequency SNVs are difficult to distinguish from technical errors, for example, due to polymerase chain reaction (PCR) amplification and sequencing protocol-related issues. It can be addressed by improving experimental methods (e.g., Primer ID (Jabara et al. 2011) or CirSeq (Lou et al. 2013)) or applying statistical models (e.g., Poisson distribution of errors (Prosperi and Salemi 2012), hidden Markov Models (Töpfer et al. 2013), or Gibbs sampling (Zagordi et al. 2011)) to the raw sequencing data. A fundamental limitation is that, in general, the quasispecies reconstruction problem does not have a unique solution if regions with low genetic diversity exist next to regions of higher diversity. For example, if two haplotypes share a genomic region which is longer than the read length, then it is not clear how to match the variants to the left with those to the right of the conserved region. Inferring the co-occurrence of mutations on the same haplotype sequence is also known as mutation phasing. In the case of viral quasispecies, it can be overcome by using long reads which cover the whole genome (but typically have higher error rates (Weirather et al. 2017)), using paired-end reads which cover segments on both sides of the conserved region, or making additional assumptions and applying heuristics in the statistical reconstruction model (e.g., by trying to match read frequencies using network flows (Baaijens et al. 2019; Astrovskaya et al. 2011)). Additionally, the true number of underlying viral strains is generally unknown and the respective strain abundances are highly variable. Inferring the number of viral haplotypes in the population is a hard problem on its own, analogous to finding the right number of clusters in unsupervised learning. While point mutations are the most common driving factor of viral quasispecies evolution, recombination, insertion, and deletion events can be more disruptive and difficult to handle in the assembly. As a consequence, there have been efforts to explicitly include them in the statistical models (e.g., recombinations in QuasiRecomb (Töpfer et al. 2013)).

4.3 OVERVIEW OF METHODOLOGICAL APPROACHES FOR GLOBAL HAPLOTYPE RECONSTRUCTION

A wide range of computational tools following diverse methodological approaches have been developed to tackle the aforementioned issues of global haplotype reconstruction, and there have been some efforts in summarizing and benchmarking this landscape (Posada-Cespedes et al. 2017; Eliseev et al. 2020).

Here, we aim to present the broad classes and categorize the global haplotype reconstruction tools which have been developed over the past decade with respect to their methodological approaches. A schematic overview of all surveyed methods is given in Figure 4.1. More information about each tool is provided in Table 4.1.

At a first stage, algorithms can be classified by whether they are reference-based, i.e., they rely on an available, high-quality reference sequence, or de-novo, meaning that they only use the provided (unaligned) sequencing reads as input. In the latter case, one may try to generate a consensus sequence of all input reads and use it as a reference in later steps, or completely forego this requirement. While reference-based approaches have generally higher reconstruction accuracies than de-novo ones, reference sequences are not always available and can introduce a bias.

Most methods use short (e.g., Illumina) sequencing reads as input, but some can also incorporate long (e.g., PacBio) reads (Artyomenko et al. 2017). Longer reads have the potential to resolve phasing issues because they cover larger parts of the genome and can thus combine information from both sides of a region of low genetic diversity, but require special care due to higher error rates. Another way of dealing with phasing is to use paired-end instead of single-end reads. Linkage of the two mates can be exploited in a similar way as long reads (Chen et al. 2018). More generally, the correction of read errors can either be done independently using specific tools (Chin et al. 2013; Salmela and Rivals 2014; Allam et al. 2015) or as part of the reconstruction. While the latter case generally makes the model more complicated and computationally expensive, modeling read errors as part of the reconstruction can be beneficial because it allows correcting for them in a more appropriate model-specific way (e.g., by constructing consensus sequences from read cliques in overlap graphs (Baaijens et al. 2017)).

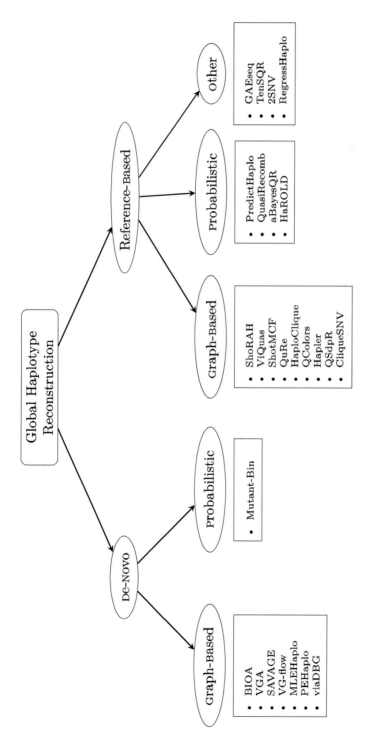

FIGURE 4.1 Schematic overview of surveyed computational tools for global haplotype reconstruction of virus populations from NGS data.

TABLE 4.1　Overview of Quasispecies Reconstruction Methods Reviewed in This Chapter

Method	Class	De-novo/ Reference-Based	Error Handling	Haplotype Reconstruction	Frequency estimation	Programming language	References
BIOA	Graph-based	De-novo	A priori correction	Max-bandwidth paths	Frequency balancing	Python 2.7	Mancuso et al. (2011)
MLEHaplo	Graph-based	De-novo	A priori correction/ graph pruning	Path scoring in graph	N/A	Perl	Malhotra et al. (2015)
PEHaplo	Graph-based	De-novo	A priori correction	Minimal path cover	N/A	Python 2.7	Chen et al. (2018)
VG-flow	Graph-based	De-novo	N/A	Contigs from, e.g., SAVAGE	Flow through variation graph	Python 3	Baaijens et al. (2019)
VGA	Graph-based	De-novo	High-fidelity sequencing protocol	Minimum vertex coloring	EM	C, Python	Mangul et al. (2014)
viaDBG	Graph-based	De-novo	A priori correction	Clique detection	N/A	C++	Freire et al. (2020)
SAVAGE	Graph-based	De-novo[a]	Constraints in graph generation	Iteratively condense read graph	N/A	C++, Python 2.7	Baaijens et al. (2017)
ViQuas	Graph-based	Reference assisted de-novo	Mutation calling	Distribution matching	Minimum frequency nodes	R, Perl	Jayasundara et al. (2015)
CliqueSNV	Graph-based	Reference-based	SNV graph	Clique detection	EM	Java	Knyazev et al. (2018)
Hapler	Graph-based	Reference-based	A priori correction/ frequency thresholding	Minimum vertex coloring	N/A	Java	O'Neil and Emrich (2011)

(Continued)

TABLE 4.1 (*Continued*) Overview of Quasispecies Reconstruction Methods Reviewed in This Chapter

Method	Class	De-novo/ Reference-Based	Error Handling	Haplotype Reconstruction	Frequency estimation	Programming language	References
HaploClique	Graph-based	Reference-based	Probabilistic sequence similarity	Max-clique merging	Normalized read counts	C++	Töpfer et al. (2014)
QColors	Graph-based	Reference-based	A priori correction	Minimum vertex coloring	N/A	C++	Huang et al. (2011)
QSdpR	Graph-based	Reference-based	N/A	Correlation clustering	Count reads per cluster	C, Python 2.7, Bash	Barik et al. (2018)
QuRe	Graph-based	Reference-based	Poisson error model	Distribution matching	Haplotype ordering	Java	Prosperi and Salemi (2012)
ShoRAH	Graph-based	Reference-based	Probabilistic clustering	Minimal path cover	EM	C++, Python 3, Perl	Zagordi et al. (2011)
ShotMCF/AmpMCF	Graph-based	Reference-based	Probabilistic read assignment	Multi-commodity flows	Normalized flow	Python 2.7	Skums et al. (2013)
ViSpA	Graph-based	Reference-based	Binomial error model	Max-bandwidth paths	EM	Java, Python 2.7, Bash	Astrovskaya et al. (2011)
2SNV	Other	Reference-based	Binomial error model	Hierarchical clustering	Uses kGEM	Scala (Java)	Artyomenko et al. (2017)
GAEseq	Other	Reference-based	Decoder of AE	Graph auto-encoder	Part of AE	C++, Python 3	Ke and Vikalo (2020)
RegressHaplo	Other	Reference-based	A priori correction	Penalized regression	Regression coefficients	R, MATLAB*	Leviyang et al. (2017)
TenSQR	Other	Reference-based	Minimize error correction score	Iterative tensor factorization	Number of reads belonging to tensor	C++, Python 3	Ahn et al. (2018)

(*Continued*)

TABLE 4.1 (*Continued*) Overview of Quasispecies Reconstruction Methods Reviewed in This Chapter

Method	Class	De-novo/ Reference-Based	Error Handling	Haplotype Reconstruction	Frequency estimation	Programming language	References
Mutant-Bin	Probabilistic	De-novo	Thresholding of low-frequency k-mers	Mixture of Poisson distributions	N/A	N/A[b]	Prabhakara et al. (2013a)
aBayesQR	Probabilistic	Reference-based	Frequency thresholding	Read clustering	Read counting	C++	Ahn and Vikalo (2018)
HaROLD	Probabilistic	Reference-based	Parameter of model	Dirichlet distribution	Parameter of model	Java	Goldstein et al. (2018)
PredictHaplo	Probabilistic	Reference-based	No local correction	Multinomial mixture model	Count reads per cluster	C++	Prabhakaran et al. (2013b)
QuasiRecomb	Probabilistic	Reference-based	Estimate position-wise sequencing error rate	Jumping HMM	Parameters of HMM	Java	Töpfer et al. (2013)

[a] Also has reference-based mode, but the de-novo mode is the main focus.
[b] Source code is not publicly available.

Viral haplotype reconstruction tools can further be divided into graph-based and probabilistic approaches, while other approaches, such as for example, deep learning (Ke and Vikalo 2020) or regressions (Leviyang et al. 2017), are less common.

The general idea of graph-based approaches is to represent some feature of the reads as nodes of a network, while some relation between these features is represented as a connection between two nodes, i.e., an edge. A common choice is the overlap graph (Myers 1995). It is, for example, used in (Baaijens et al. 2017; Töpfer et al. 2014). In this representation, nodes correspond to reads and a directed edge is created between node v and w, if a suffix of v matches a prefix of w for a sufficient length and quality score. Such graphs can then be simplified by, for example, transitive edge removal. Haplotypes can then be identified by searching for clusters of nodes in the graph. Overlap graphs can also be extended by endowing each edge with a weight. In particular, the edge weight can reflect the number of matches and mismatches on overlapping segments (Barik et al. 2018). This extension affects the node clustering procedure by making the cluster structure more dependent on high-quality overlaps. It is also possible to explicitly model paired-end reads during the creation of the overlap graph (Chen et al. 2018). The paired-end overlap graph contains the edges of the original overlap graph and, in addition, edges which connect nodes whose reads form paired-end reads. This procedure is beneficial during graph pruning and pathfinding. Another formulation which became popular in the context of genome assembly is the de Bruijn graph (De Bruijn 1946; Compeau et al. 2011). Its nodes correspond to k-mers obtained from all reads and its edges denote their overlap. Overlap graphs are typically more computationally demanding to build, while de Bruijn graphs are more susceptible to read errors. See Rizzi et al. (2019) for a more in-depth comparison. Variation graphs and their extension to flow variation graphs are particularly useful for estimating haplotype frequencies (Garrison et al. 2018). Their nodes correspond to substrings of the input reads and edges indicate that a concatenation of these substrings would result in another substring of the input reads. In addition, a variation graph stores a set of paths that correspond to the input reads, as well as path abundances. This approach can be used to efficiently find optimal flows through the graph which then correspond to global haplotypes (Baaijens et al. 2019). Conflict graphs are complementary to overlap graphs in that they model each read as a node but only create an edge if the reads overlap but conflict (i.e., mismatch) in some alleles (O'Neil and Emrich 2011; Huang et al. 2011;

Mangul et al. 2014). They allow, for example, using a minimum coloring approach to identify haplotypes.

The general idea of probabilistic approaches is to model reads and their error profiles in a statistical model, estimate the model parameters, and then read off haplotype sequences and frequencies. Hidden Markov models (HMMs) generate haplotypes from the base alphabet and transition probabilities (Töpfer et al. 2013). The structure of HMMs makes it possible to explicitly account for recombination events in the model. Dirichlet process mixture models (DPMMs) can be used in a Bayesian formulation of the global haplotype reconstruction problem (Prabhakaran et al. 2013b). They require few input parameters, are computationally efficient, and alleviate potential local minima issues which can arise when employing expectation-maximization (EM) approaches compared to other Bayesian approaches (Zagordi et al. 2010).

Several other and diverse approaches exist. Graph auto-encoders can be used to formulate the haplotype reconstruction problem as first encoding undirected bipartite read SNV graphs and then retrieving the haplotypes by decoding components from the latent space (Ke and Vikalo 2020). Here, the low-dimensional latent space tries to enforce a sparse set of reconstructed haplotypes. An iterative tensor factorization approach represents input reads as structured sparse binary tensors (Ahn et al. 2018). It overcomes the problem of highly variable haplotype frequencies by iterating a two-step process, which factorized the tensor and reconstructs the most abundant strain first, and then removes all reads belonging to this strain. In a penalized regression approach, the penalty term aims at reconstructing a sparse set of haplotypes which have been generated from a read graph (Leviyang et al. 2017).

4.4 CONCLUSIONS

In summary, global haplotype reconstruction methods pose a promising avenue for improving our understanding of viral evolutionary dynamics, but not all technical issues have been fully solved. On the methodological side, theoretical limitations and data processing issues need to be addressed before clinically significant interpretations can be made. However, due to a lack of standardized benchmarking comparisons between competing software, selecting the appropriate tool for each application remains difficult. Recently, efforts have been made to tackle this problem (Eliseev et al. 2020). Conducting an unbiased and thorough benchmark is challenging. Each tool may depend on ancient compiler versions, peculiar software dependencies, or specific operating systems. Furthermore, too few real datasets with a known ground truth are available and synthetic datasets have biases and may be unrealistic not reflecting viral evolution and epidemiology properly. At the same time, choosing the best performance metric of a haplotype reconstruction is also not straightforward and depends on the focus of the study and the question it tries to address. It will furthermore be interesting to see new methods combining long and short reads to make use of improved mutation phasing capabilities and lower error rates.

Various software engineering aspects have an impact on how easy a tool is to install and use, and thus how helpful it can be in practice for global haplotype reconstruction efforts of other researchers. Making the software (and its source) easily available, for example, by uploading it to public Git repositories, is crucial so that others can use it. Preferably, the software should be available as a Conda package (Anaconda Inc. 2020) or Docker image (Merkel 2014). Nonetheless, it will be exciting to see how better quasispecies reconstruction methods will be influential in the medical field.

REFERENCES

Ahn, S. and Vikalo, H. 2018. aBayesQR: A Bayesian method for reconstruction of viral populations characterized by low diversity. *Journal of Computational Biology* 25(7), pp. 637–648.

Ahn, S., Ke, Z. and Vikalo, H. 2018. Viral quasispecies reconstruction via tensor factorization with successive read removal. *Bioinformatics* 34(13), pp. i23–i31.

Allam, A., Kalnis, P. and Solovyev, V. 2015. Karect: Accurate correction of substitution, insertion and deletion errors for next-generation sequencing data. *Bioinformatics* 31(21), pp. 3421–3428.

Anaconda Inc. 2020. Anaconda software distribution. Computer software. Anaconda Documentation.

Artyomenko, A., Wu, N.C., Mangul, S., Eskin, E., Sun, R. and Zelikovsky, A. 2017. Long single-molecule reads can resolve the complexity of the influenza virus composed of rare, closely related mutant variants. *Journal of Computational Biology* 24(6), pp. 558–570.

Astrovskaya, I., Tork, B., Mangul, S., et al. 2011. Inferring viral quasispecies spectra from 454 pyrosequencing reads. *BMC Bioinformatics* 12(Suppl 6), p. S1.

Baaijens, J.A., Aabidine, A.Z.E., Rivals, E. and Schönhuth, A. 2017. De novo assembly of viral quasispecies using overlap graphs. *Genome Research* 27(5), pp. 835–848.

Baaijens, J.A., Stougie, L. and Schönhuth, A. 2019. Strain-aware assembly of genomes from mixed samples using variation graphs. *BioRxiv*.

Barik, S., Das, S. and Vikalo, H. 2018. QSdpR: Viral quasispecies reconstruction via correlation clustering. *Genomics* 110(6), pp. 375–381.

Beerenwinkel, N. and Zagordi, O. 2011. Ultra-deep sequencing for the analysis of viral populations. *Current Opinion in Virology* 1(5), pp. 413–418.

Beerenwinkel, N., Günthard, H.F., Roth, V. and Metzner, K.J. 2012. Challenges and opportunities in estimating viral genetic diversity from next-generation sequencing data. *Frontiers in Microbiology* 3, p. 329.

Chen, J., Zhao, Y. and Sun, Y. 2018. De novo haplotype reconstruction in viral quasispecies using paired-end read guided path finding. *Bioinformatics* 34(17), pp. 2927–2935.

Chin, C.-S., Alexander, D.H., Marks, P., et al. 2013. Nonhybrid, finished microbial genome assemblies from long-read SMRT sequencing data. *Nature Methods* 10(6), pp. 563–569.

Compeau, P.E.C., Pevzner, P.A. and Tesler, G. 2011. How to apply de Bruijn graphs to genome assembly. *Nature Biotechnology* 29(11), pp. 987–991.

De Bruijn, N.G. 1946. A combinatorial problem. *Proceedings of the Section of Sciences of the Koninklijke Nederlandse Akademie van Wetenschappen te Amsterdam* 49(7), pp. 758–764.

Dong, E., Du, H. and Gardner, L. 2020. An interactive web-based dashboard to track COVID-19 in real time. *The Lancet Infectious Diseases* 20(5), pp. 533–534.

Duffy, S., Shackelton, L.A. and Holmes, E.C. 2008. Rates of evolutionary change in viruses: Patterns and determinants. *Nature Reviews Genetics* 9(4), pp. 267–276.

Eigen, M. and Schuster, P. 1977. A principle of natural self-organization. *Die Naturwissenschaften* 64(11), pp. 541–565.

Eliseev, A., Gibson, K.M., Avdeyev, P., et al. 2020. Evaluation of haplotype callers for next-generation sequencing of viruses. *Infection, Genetics and Evolution* 82, p. 104277.

Freire, B., Ladra, S., Paramá, J. and Salmela, L. 2020. Inference of viral quasispecies with a paired de Bruijn graph. *Bioinformatics*. https://academic.oup.com/bioinformatics/advance-article/doi/10.1093/bioinformatics/btaa782/5905473

Garrison, E., Sirén, J., Novak, A.M., et al. 2018. Variation graph toolkit improves read mapping by representing genetic variation in the reference. *Nature Biotechnology* 36(9), pp. 875–879.

Gaschen, B., Taylor, J., Yusim, K., et al. 2002. Diversity considerations in HIV-1 vaccine selection. *Science* 296(5577), pp. 2354–2360.

Goldstein, R.A., Tamuri, A.U., Roy, S. and Breuer, J. 2018. Haplotype assignment of virus NGS data using co-variation of variant frequencies. *BioRxiv*.

Huang, A., Kantor, R., DeLong, A., Schreier, L. and Istrail, S. 2011. QColors: An algorithm for conservative viral quasispecies reconstruction from short and non-contiguous next generation sequencing reads. *In Silico Biology* 11(5–6), pp. 193–201.

Jabara, C.B., Jones, C.D., Roach, J., Anderson, J.A. and Swanstrom, R. 2011. Accurate sampling and deep sequencing of the HIV-1 protease gene using a Primer ID. *Proceedings of the National Academy of Sciences of the United States of America* 108(50), pp. 20166–20171.

Jayasundara, D., Saeed, I., Maheswararajah, S., Chang, B.C., Tang, S.L. and Halgamuge, S.K. 2015. ViQuaS: An improved reconstruction pipeline for viral quasispecies spectra generated by next-generation sequencing. *Bioinformatics* 31(6), pp. 886–896.

Johnson, J.A., Li, J.-F., Wei, X., et al. 2008. Minority HIV-1 drug resistance mutations are present in antiretroviral treatment-naïve populations and associate with reduced treatment efficacy. *PLoS Medicine* 5(7), p. e158.

Ke, Z. and Vikalo, H. 2020. A graph auto-encoder for haplotype assembly and viral quasispecies reconstruction. *Proceedings of the AAAI Conference on Artificial Intelligence* 34(01), pp. 719–726.

Knyazev, S., Tsyvina, V., Melnyk, A., et al. 2018. CliqueSNV: Scalable reconstruction of intra-host viral populations from NGS reads. *BioRxiv.*

Leviyang, S., Griva, I., Ita, S. and Johnson, W.E. 2017. A penalized regression approach to haplotype reconstruction of viral populations arising in early HIV/SIV infection. *Bioinformatics* 33(16), pp. 2455–2463.

Lou, D.I., Hussmann, J.A., McBee, R.M., et al. 2013. High-throughput DNA sequencing errors are reduced by orders of magnitude using circle sequencing. *Proceedings of the National Academy of Sciences of the United States of America* 110(49), pp. 19872–19877.

Malhotra, R., Wu, M.M.S., Rodrigo, A., Poss, M. and Acharya, R. 2015. Maximum Likelihood de novo reconstruction of viral populations using paired end sequencing data. arXiv.

Mancuso, N., Tork, B., Skums, P., Mandoiu, I. and Zelikovsky, A. 2011. Viral quasispecies reconstruction from amplicon 454 pyrosequencing reads. In: *2011 IEEE International Conference on Bioinformatics and Biomedicine Workshops (BIBMW)*. Atlanta, GA, IEEE, pp. 94–101.

Mangul, S., Wu, N.C., Mancuso, N., Zelikovsky, A., Sun, R. and Eskin, E. 2014. Accurate viral population assembly from ultra-deep sequencing data. *Bioinformatics* 30(12), pp. i329–i337.

Marz, M., Beerenwinkel, N., Drosten, C., et al. 2014. Challenges in RNA virus bioinformatics. *Bioinformatics* 30(13), pp. 1793–1799.

Merkel, D. 2014. Docker: Lightweight Linux containers for consistent development and deployment. *Linux Journal* 2014(239), p. 2.

Myers, E.W. 1995. Toward simplifying and accurately formulating fragment assembly. *Journal of Computational Biology* 2(2), pp. 275–290.

Nowak, M.A., Anderson, R.M., McLean, A.R., Wolfs, T.F., Goudsmit, J. and May, R.M. 1991. Antigenic diversity thresholds and the development of AIDS. *Science* 254(5034), pp. 963–969.

O'Neil, S.T. and Emrich, S.J. 2011. Robust haplotype reconstruction of eukaryotic read data with Hapler. In: *2011 IEEE 1st International Conference on Computational Advances in Bio and Medical Sciences (ICCABS)*. Orlando, FL, IEEE, pp. 141–146.

Posada-Cespedes, S., Seifert, D. and Beerenwinkel, N. 2017. Recent advances in inferring viral diversity from high-throughput sequencing data. *Virus Research* 239, pp. 17–32.

Prabhakara, S., Malhotra, R., Acharya, R. and Poss, M. 2013a. Mutant-bin: Unsupervised haplotype estimation of viral population diversity without reference genome. *Journal of Computational Biology* 20(6), pp. 453–463.

Prabhakaran, S., Rey, M., Zagordi, O., Beerenwinkel, N. and Roth, V. 2013b. HIV haplotype inference using a propagating Dirichlet process mixture model. *IEEE/ACM Transactions on Computational Biology and Bioinformatics* 11(1), pp. 182–192.

Prosperi, M.C.F. and Salemi, M. 2012. QuRe: Software for viral quasispecies reconstruction from next-generation sequencing data. *Bioinformatics* 28(1), pp. 132–133.

Rizzi, R., Beretta, S., Patterson, M., et al. 2019. Overlap graphs and de Bruijn graphs: Data structures for de novo genome assembly in the big data era. *Quantitative Biology* 7(4), pp. 278–292.

Salmela, L. and Rivals, E. 2014. LoRDEC: Accurate and efficient long read error correction. *Bioinformatics* 30(24), pp. 3506–3514.

Seifert, D. and Beerenwinkel, N. 2016. Estimating fitness of viral quasispecies from next-generation sequencing data. *Current Topics in Microbiology and Immunology* 392, pp. 181–200.

Shendure, J., Balasubramanian, S., Church, G.M., et al. 2017. DNA sequencing at 40: Past, present and future. *Nature* 550(7676), pp. 345–353.

Skums, P., Mancuso, N., Artyomenko, A., et al. 2013. Reconstruction of viral population structure from next-generation sequencing data using multi-commodity flows. *BMC Bioinformatics* 14(Suppl 9), p. S2.

Töpfer, A. and Beerenwinkel, N. 2016. Probabilistic viral quasispecies assembly. In: Măndoiu, I. and Zelikovsky, A. eds. *Computational Methods for Next Generation Sequencing Data Analysis*. Hoboken, NJ: John Wiley & Sons, Inc., pp. 355–381.

Töpfer, A., Zagordi, O., Prabhakaran, S., Roth, V., Halperin, E. and Beerenwinkel, N. 2013. Probabilistic inference of viral quasispecies subject to recombination. *Journal of Computational Biology* 20(2), pp. 113–123.

Töpfer, A., Marschall, T., Bull, R.A., Luciani, F., Schönhuth, A. and Beerenwinkel, N. 2014. Viral quasispecies assembly via maximal clique enumeration. *PLoS Computational Biology* 10(3), p. e1003515.

Tsibris, A.M.N., Korber, B., Arnaout, R., et al. 2009. Quantitative deep sequencing reveals dynamic HIV-1 escape and large population shifts during CCR5 antagonist therapy in vivo. *PLoS One* 4(5), p. e5683.

Weirather, J.L., de Cesare, M., Wang, Y., et al. 2017. Comprehensive comparison of Pacific Biosciences and Oxford Nanopore Technologies and their applications to transcriptome analysis. [version 2; peer review: 2 approved]. *F1000Research* 6, p. 100.

Zagordi, O., Geyrhofer, L., Roth, V. and Beerenwinkel, N. 2010. Deep sequencing of a genetically heterogeneous sample: Local haplotype reconstruction and read error correction. *Journal of Computational Biology* 17(3), pp. 417–428.

Zagordi, O., Bhattacharya, A., Eriksson, N. and Beerenwinkel, N. 2011. ShoRAH: Estimating the genetic diversity of a mixed sample from next-generation sequencing data. *BMC Bioinformatics* 12, p. 119.

CHAPTER 5

Functional RNA Structures in the 3′ UTR of Mosquito-Borne Flaviviruses

Michael T. Wolfinger, Roman Ochsenreiter, and Ivo L. Hofacker

University of Vienna

CONTENTS

5.1 INTRODUCTION

Flaviviruses (genus *Flavivirus*) are enveloped, non-segmented single-stranded (+)-sense RNA viruses of ~50 nm diameter with icosahedral or spherical geometries. They belong to the *Flaviviridae* family, which also includes the genera *Hepacivirus*, *Pegivirus*, and *Pestivirus*. Flaviviruses are distributed worldwide, with certain species found only in endemic or epidemic areas, and can infect a wide variety of vertebrate and invertebrate species, rendering them a global economic burden and major health threat. Most flaviviruses are zoonotic arthropod-borne viruses (arboviruses) with ancestral transmission cycles in wildlife. Humans are typically infected via direct spillover (enzootic cycle), amplification in domesticated animals followed by spillover to humans (epizootic cycle), or a human–mosquito–human cycle, often in urban environments (urban epidemic cycle) [1]. While these transmission cycles involve hematophagous arthropod vectors, predominantly of the *Aedes* and *Culex* genera, the transmission mechanism has not been elucidated for all flaviviruses. This is reflected in the classification into groups that represent ecological host/vector associations (Figure 5.1): Mosquito-borne flaviviruses (MBFVs) and tick-borne flaviviruses (TBFVs) spread between invertebrate vectors (mosquitoes and ticks) and vertebrate hosts (mainly mammals and birds) and comprise the majority of all known flaviviruses. They are also referred to as vertebrate-infecting flaviviruses (VIFs) in the literature. Insect-specific flaviviruses (ISFs) only replicate in mosquitoes and split into two groups: classic insect-specific flaviviruses (cISFs) and dual-host affiliated insect-specific flaviviruses (dISFs) [2]. The first group exclusively infects mosquitoes and forms a monophyletic group within the flavivirus phylogenetic tree, while the second group comprises viruses in separate clades that are more closely related to mosquito/vertebrate flaviviruses without the capability to infect vertebrate cells. Contrary, no-known-vector flaviviruses (NKVs) comprise a phylogenetically and ecologically diverse set of viruses that have only been found in rodents and bats, without causing disease or high viremia in these animals [3], although human infection and disease have been reported [4]. Importantly, no arthropod vector has been identified so far for these viruses. The recent discovery of novel crustacean flaviviruses that share ancestral roots with terrestrial vector-borne flaviviruses and the characterization of cephalopod flaviviruses that are the most divergent of all flaviviruses currently known highlight the importance of these marine

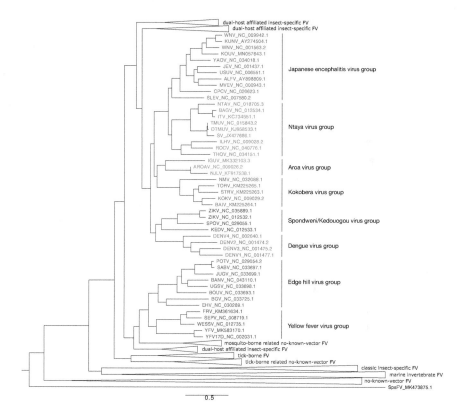

FIGURE 5.1   Maximum likelihood tree of the genus *Flavivirus*, computed from a **MAFFT** multiple sequence alignment of complete polyprotein nucleotide sequences with **iqtree**. Serologically distinguishable groups (most of them are defined by the International Committee on Taxonomy of Viruses (ICTV)) are in good agreement with sequence-based phylogenetic clustering and high-lighted in color. Viral species are abbreviated together with the NCBI accession number that was used for inferring the phylogeny. **refseq** isolates were used whenever possible. For species without **refseq** annotation, the longest complete genome from **genbank** for this species was selected as representative sequence. Non-mosquito-borne flavivirus clades have been collapsed to contextualize their positioning in the flavivirus phylogeny. Abbreviations follow the nomenclature discussed in the main text. Southern pygmy squid flavivirus (SpsFV, bottom) is a marine invertebrate flavivirus with the largest phylogenetic distance to all other virus taxa considered here. The tree has been rendered with **FigTree**.

invertebrate flaviviruses (MIFs) in our understanding of flavivirus vertebrate-invertebrate association and evolution [5].

Of the above groups, the vertebrate-infecting flaviviruses certainly represent a major challenge for public health systems, as they comprise re-emerging pathogens like Dengue virus (DENV), West Nile virus (WNV), Japanese encephalitis virus (JEV), and tick-borne encephalitis virus (TBEV) as well as recently emerging viruses like Zika virus (ZIKV) and Usutu virus (USUV). All of these can cause disease in humans with different clinical manifestations ranging from asymptomatic to febrile illness, arthralgia, rash, and hemorrhagic fever. Importantly, several mosquito-borne flaviviruses are neurotropic and can induce severe pathology of the central nervous system, including meningitis, encephalitis, and fetal microcephaly [6]. On the other side, there are only experimental antiviral drugs to treat flavivirus infections, and the number of available vaccines is limited. The most prominent is the YFV vaccine based on the live attenuated YFV-17D strain developed by Max Theiler and collaborators in 1937. Interestingly, this vaccine is still working after more than 80 years and has been used to prevent yellow fever infections in hundreds of millions of people. Besides that, inactivated JEV and TBEV vaccines are available for use in humans, and an inactivated WNV vaccine is available for use in animals [7].

In addition to the classification by ecological groups, the International Committee on Taxonomy of Viruses (ICTV) proposes a more fine-grained association of flaviviruses into antigenic complexes based on serological evidence. For the mosquito-borne flaviviruses, this results in the establishment of eight distinguishable groups [8]: Japanese encephalitis virus group (JEVG), Ntaya virus group (NTAVG), Aroa virus group (AROAVG), Kokobera virus group (KOKVG), Spondweni/Kedougou virus group (SPOVG), Dengue virus group (DENVG), Edge hill virus group (EHVG), and Yellow fever virus group (YFVG). The maximum likelihood tree in Figure 5.1 illustrates that this classification is in good agreement with sequence-based molecular phylogenetics, which can be explained by the common genome organization shared by all flaviviruses. Flavivirus genomes consist of a single 5′-capped, non-polyadenylated RNA of ~10–12 kilobases in length, referred to as gRNA. The gRNA encodes a single open reading frame (ORF) of ~3,400 codons, which is flanked by highly structured untranslated regions (UTRs) [9]. Upon translation of the ORF, a polyprotein is produced which is processed by viral and cellular enzymes, yielding structural (C, prM, E) and nonstructural (NS1, NS2A, NS2B, NS3,

NS4A, 2K, NS4B, NS5) proteins [10]. Conversely, both flavivirus UTRs are crucially involved in the regulation of the viral life cycle, thereby mediating processes such as genome cyclization, viral replication, packaging, and immune response [11,12]. Although the UTRs show very little primary sequence conservation across the different ecological flavivirus groups, RNA structure conservation is eminent [13,14]. This lines up flaviviruses with other RNA viruses that have evolved functional RNAs [15], such as Hepatitis C virus (HCV), Picorna- and Hepaciviruses with their well-studied 5′ UTR internal ribosome-entry (IRES) sites [16,17], alphaviruses that have predicted structures of unknown function in their 3′ UTR [18], or coronaviruses with their cis-acting UTR elements [19–22].

5.2 FLAVIVIRUS 3′ UTR BACKGROUND

Biologically functional RNA elements often rely on a specific fold, which is manifested in a particular, evolutionarily conserved structure. Rather than maintaining the primary sequence, nature implements conservation typically at the level of RNA secondary structures, rendering the task of finding conserved elements essentially a variant of the in silico RNA structure prediction problem. Structural conservation is often achieved through compensatory mutations, where the combination of two point mutations results in maintenance of complementarity, as in a mutation AU → GC. Homology search approaches build on this evolutionary trait that is also known as covariation. We have recently employed this methodology to propose a varied landscape of RNA structure conservation in the 3′ UTR of tick-borne, insect-specific, and no-known-vector flaviviruses based on comparative genomics screens [23]. Here we follow up on our previous investigations and establish a comprehensive map of functional RNAs in the 3′ UTRs of mosquito-borne flaviviruses.

A hallmark of flavivirus biology that has been elucidated over the last decade is their ability to actively dysregulate the host mRNA degradation machinery. They achieve this by tolerating the degradation of parts of their genome by endogenous host enzymes, resulting in the accumulation of viral long noncoding RNA (lncRNAs) species in infected cells. These lncRNAs are referred to as subgenomic flaviviral RNAs (sfRNAs) [24] and are stable decay intermediates resulting from incomplete degradation of viral gRNA by host exoribonuclease activity [25]. sfRNAs eventually dysregulate cellular function to evade the host antiviral response and promoting viral infection [26]; however, there are major differences in the sfRNA characteristics between different flavivirus groups. Evidence of sfRNA production following incomplete degradation by the 5′ → 3′ exoribonuclease Xrn1, an enzyme associated with the cell's RNA turnover machinery [27], has been observed in mosquito-borne [28,29], tick-borne [30], no-known-vector [31], and dual-host affiliated insect-specific flaviviruses [32]. Mechanistically, sfRNAs are produced by halting Xrn1 at evolutionary highly conserved exoribonuclease-resistant RNA (xrRNA) elements that exhibit an ion-dependent mechanical anisotropy [33]. Stalling at xrRNAs prohibits further Xrn1 progression toward the viral 3′ end and confers quantitative protection of downstream nucleotides [34].

Mosquito-borne flaviviruses have typically more than one xrRNA element, each with different exoribonuclease stalling capacity, thereby enabling the production of sfRNAs of different lengths [35]. Many of the conserved structures associated with xrRNA functionality have been computationally predicted prior to experimental validation [17,36–38]. Specifically, the so-called stem-loop (SL) and dumbbell (DB) elements found in the 3′ UTR of many flaviviruses have been related to quantitative protection of downstream virus RNA against Xrn1 degradation. Figure 5.2 shows the complete Usutu virus 3′ UTR (665 nt) with all its predicted RNA secondary structure elements, showcasing RNA elements that confer Xrn1-resistance.

The 3′ UTR of mosquito-borne flavivirus genomes comprises three autonomously folded regions, i.e., domains I-III, that contain one or two copies of conserved RNA elements with distinct functional associations, as shown in Figure 5.2. Domain I, which is located immediately downstream of the NS5 protein stop codon, manifests as the most variable region, containing a set of simple and branched SL elements. Importantly, due to the intrinsic volatility of this region, there is no consistent naming scheme for these elements in the literature. Two of the SL elements in domain I (designated as SL-II and SL-IV here) stand out, as structural homologs can be found in all flaviviruses [23,39]. They are associated with the formation of a three-way junction structure that can form a transient pseudoknot by base-pairing of the apical loop with nucleotides located immediately downstream of the stem-loop element. The resulting structure is a tightly packed, mechanically stable xrRNA. Between the two xrRNA-associated SL-II and SL-IV elements, a bulged stem-loop element (SL-III) of ~70 nt length that is interspersed with interior loops is found in species of the Japanese encephalitis virus group and Ntaya virus group as well as in several species of the dual-host affiliated insect-specific flaviviruses [32]. Although the biological function of this element is unknown, evolutionary

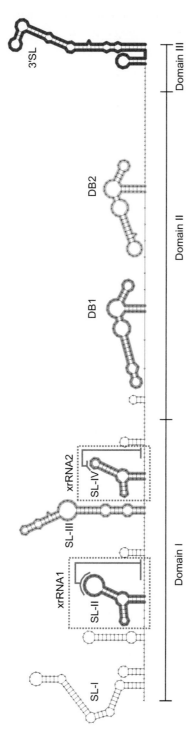

FIGURE 5.2   3' UTR organization of mosquito-borne flaviviruses, exemplarily shown for Usutu virus, a member of the Japanese encephalitis group. The three structurally and functionally independent domains contain duplicated stem-loop (SL) and dumbbell (DB) elements as well as a terminal 3' stem-loop (3' SL) structure. SL elements in domain I are numbered from the beginning of the 3' UTR. Elements that are well conserved across multiple species are highlighted in color. The duplicated SL-II and SL-IV elements fold into a three-way junction structure that can form a transient pseudoknot with sequence regions downstream of the SL elements (depicted in grey), thereby producing exoribonuclease-resistant RNAs xrRNA1 and xrRNA2 (highlighted in boxes).

conservation across different ecological flavivirus groups is intriguing and could be related to specific vertebrate-invertebrate associations of these viruses. Domain II contains either one or two DB elements, some of which have been reported to also form pseudoknot interactions with downstream regions [40]. These structures bear some exoribonuclease-stalling activity, although to a weaker extent than the upstream SL-derived xrRNAs [34]. An interesting property of the DB elements is the high degree of primary sequence conservation in the distal part of the central multi-loop and the adjacent hairpin loop [41], which can be explained by overlap of the DB elements with cyclization sequences [42] that mediate the formation of long-range RNA–RNA interactions required for virus replication [43,44]. Finally, domain III contains a small hairpin (sHP) and a larger bulged SL structure, termed 3′ stem-loop (3′ SL), which is thermodynamically particularly stable and conserved in all flaviviruses. Host factors acting as chaperones can destabilize the 3′ SL element, thereby making cyclization sequences accessible and mediating the process of pan-handle formation via long-range RNA–RNA interactions between the genomic 5′ and 3′ ends [45].

5.3 MATERIALS AND METHODS

The computational prediction of evolutionarily conserved RNA structures in viruses has become an attractive research topic over the last two decades, particularly with the increasing availability of viral genome and metagenome data. Viral genome data, including sequences and annotation presented here, was downloaded from the public National Center for Biotechnology Information (NCBI) **refseq** (https://www.ncbi.nlm.nih.gov/refseq/) and **genbank** (https://www.ncbi.nlm.nih.gov/genbank/) databases on 26 May 2020. We obtained complete viral genomes under taxonomy ID 11051 (genus *Flavivirus*) and filtered for mosquito-borne flavivirus species listed in Table 5.1. For phylogeny reconstruction, we aimed at selecting **refseq** entries as representatives for each species. In case **refseq** isolates were not available for a particular species, we selected one of the longest complete genomes from the **genbank** set as representative sequence. A maximum-likelihood phylogeny of all species in the genus *Flavivirus* based on a **MAFFT** [46] multiple sequence alignment of complete coding region nucleotide sequences has been inferred with **iq-tree** [47] using the SYM+R10 substitution model (Figure 5.1).

Detection and characterization of evolutionarily conserved RNAs in MBFV 3′ UTRs follow the methodology described recently for other

TABLE 5.1   Mosquito-Borne Flavivirus Genomes Considered in This Contribution

Accession Number[a]	Acronym	Scientific Name	3′ UTR Length (nt)	Isolates
AY898809.1	ALFV	Alfuy virus	560	1
NC_009026.2	AROAV	Aroa virus	421	1
NC_012534.1	BAGV	Bagaza virus	566	19
KM225264.1	BAIV	Bainyik virus	371	1
NC_043110.1	BANV	Banzi virus	NA	NA
NC_033725.1	BGV	Bamaga virus	NA	NA
NC_033693.1	BOUV	Bouboui virus	NA	NA
LN849009.1	CPCV	Cacipacore virus	573	1
NC_001477.1	DENV1	Dengue virus 1	462	2,210
NC_001474.2	DENV2	Dengue virus 2	451	1,631
NC_001475.2	DENV3	Dengue virus 3	440	1,028
NC_002640.1	DENV4	Dengue virus 4	384	260
KJ958533.1	DTMUV	Duck Tembusu virus	692	19
NC_030289.1	EHV	Edge Hill virus	NA	NA
KM361634.1	FRV	Fitzroy River virus	472	1
AY632538.4	IGUV	Iguape virus	567	5
NC_009028.2	ILHV	Ilheus virus	388	3
KC734551.1	ITV	Israel turkey meningoencephalomyelitis virus	434	5
NC_001437.1	JEV	Japanese encephalitis virus	582	289
NC_033699.1	JUGV	Jugra virus	NA	NA
NC_012533.1	KEDV	Kedougou virus	390	2
NC_009029.2	KOKV	Kokobera virus	558	2
MN057643.1	KOUV	Koutango virus	551	1
AY274504.1	KUNV	Kunjin virus	624	35
NC_000943.1	MVEV	Murray Valley encephalitis virus	614	18
KF917538.1	NJLV	Naranjal virus	NA	NA
NC_032088.1	NMV	New Mapoon virus	546	2
NC_018705.3	NTAV	Ntaya virus	565	2
NC_029054.2	POTV	Potiskum virus	NA	NA
NC_040776.1	ROCV	Rocio virus	424	3
NC_033697.1	SABV	Saboya virus	NA	NA
NC_008719.1	SEPV	Sepik virus	459	2
NC_007580.2	SLEV	Saint Louis encephalitis virus	549	44
SA-AR[b]	SPOV	Spondweni virus	338	1
KM225263.1	STRV	Stratford virus	233	1
JX477686.1	SV	Sitiawan virus	NA	NA
NC_034151.1	THOV	T'Ho virus	556	2
NC_015843.2	TMUV	Tembusu virus	618	74

(*Continued*)

TABLE 5.1 (*Continued*) Mosquito-Borne Flavivirus Genomes Considered in This Contribution

Accession Number[a]	Acronym	Scientific Name	3′ UTR Length (nt)	Isolates
KM225265.1	TORV	Torres virus	414	1
NC_033698.1	UGSV	Uganda S virus	NA	NA
NC_006551.1	USUV	Usutu virus	665	287
NC_012735.1	WESSV	Wesselsbron virus	478	7
NC_009942.1	WNV	West Nile virus (lineage 1)	631	1,920[c]
NC_001563.2	WNV	West Nile virus (lineage 2)	573	1,920[c]
NC_034018.1	YAOV	Yaounde virus	NA	NA
MK583170.1	YFV	Yellow fever virus	657	196
NC_002031.1	YFV17D	Yellow fever virus 17D	508	23
NC_035889.1	ZIKV	Zika virus (Asian/American lineage)	429	801[c]
NC_012532.1	ZIKV	Zika virus (African lineage)	428	801[c]

The 3′ UTR length is listed for each isolate.

[a] Representative accession number from the refseq database. Whenever a refseq genome was not available, the isolate with the longest 3′ UTR from the genbank database was selected as representative species.

[b] SPOV 3′ UTR data taken from Ref. [41].

[c] Total number of available isolates for this virus.

NA: 3′ UTR partial or not available.

flaviviruses [23]. Our approach is centered around finding structurally homologous RNA elements in phylogenetically narrow subgroups by means of RNA family models. We use **infernal** [48] to transform structural RNA alignments into covariance models [49], i.e., statistical models of RNA structure that extend classic Hidden Markov Models (HMMs) to simultaneously represent sequence and secondary structure. Covariance models can then be used to find homologous RNAs in large sequence databases. Building on stochastic context-free grammars, a covariance model returns for each hit a bit score, as well as an E-value to assess hit quality.

Here we construct for each MBFV antigenic complex structural multiple sequence alignments of all available 3′ UTR sequences from the **refseq** set with [50]. Stringent comparison of the predicted consensus structures with known functional RNAs, such as those listed in Rfam [51], allows for exact localization and realignment of conserved elements. In parallel, we compute locally stable RNA secondary structures from structurally aligned UTR regions with **RNALalifold** [52] from the **ViennaRNA** package [53]. Covariance models are then built for every conserved RNA structure in each antigenic complex, resulting in a total of 37 covariance models that are subsequently used to perform a comprehensive screen in all **genbank** sequences studied here. As covariance models constructed this way are highly specific for a particular copy of a conserved structural element within an antigenic group, this methodology can discriminate paralogous copies, i.e., SL-II/SL-IV or DB1/DB2, and infer structural proximity among conserved elements in a quantitative manner. We use this approach to unambiguously characterize individual copies as SL-II/SL-IV or DB1/DB2. Structural alignments of all conserved elements presented in this contribution are available at https://github.com/mtw/viRNA.

To learn more about the evolutionary traits associated with the presence of single or duplicated functional elements in the 3′ UTRs, we perform a quantitative comparison of covariance models based on **CMCompare** [54]. This tool computes for a pair of covariance models a link score that exhibits high values if a sequence exists that scores well in both models simultaneously. Analysis of the link scores computed from pairwise covariance model comparisons yields valuable insight into the association of conserved functional elements. In particular, this strategy allows for a fine-grained classification of elements that are found in varying copy numbers in some viruses and reveals elements that are unique to certain species or antigenic groups.

5.4 RESULTS

Detailed experimental studies over the last decades have revealed the functional association of flavivirus RNAs, in particular evolutionarily conserved structural elements in the UTRs of these viruses, with pathogenesis (reviewed in Ref. [55]), leading to a broad understanding of the biological roles and implications of viral non-coding RNAs (ncRNAs). While many studies focus on examples in individual viruses, thereby elucidating properties of single sequences, a unifying view of the complex flavivirus 3′ UTR architecture and the evolutionary association among homologous RNA elements are only beginning to be understood. Following up on our previous investigations in tick-borne, insect-specific, and no-known-vector flaviviruses [23], we present here a complete picture of the evolutionary conservation of functional RNAs in the 3′ UTR of mosquito-borne flaviviruses. Pursuing a bipartite comparative genomics strategy for elucidating conserved viral RNAs, i.e., building on previously known xrRNA, DB, and 3′ SL elements in specific isolates, as well as characterizing hitherto understudied elements de novo allows us to draw a comprehensive map of structurally homologous RNAs in this ecologic group of viruses. We constrain

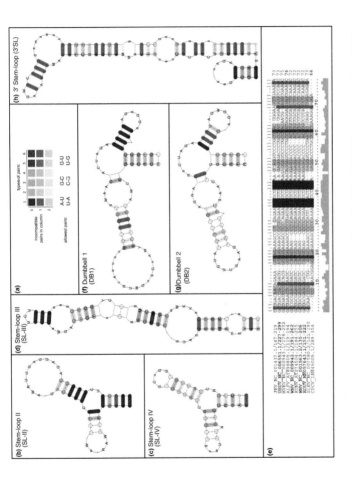

FIGURE 5.3 Consensus structures of conserved RNA elements in the 3' UTR of JEVG virus species. (a) Base-pair coloring in this figure indicates different levels of covariation observed at the corresponding columns in the underlying alignment, ranging from red (no covariation) over ocher, green, and blue to violet (representing all six base-pair combinations). (b) and (c) Stem-loop elements II and IV fold into canonical three-way junction structures and constitute building blocks for xrRNAs. (d) Stem-loop III is a highly conserved stem-loop structure interspersed with several interior loops. Biological function of this element is unknown. (e) The structural alignment of SL-III, underlying the consensus structure (d) is shown as an example. Gray bars below the alignment indicate for each column the level of primary sequence conservation. (f) and (g) Dumbbell elements 1 and 2. Sequence-level nucleotide conservation in the distal stem-loop of all species is due to overlap with cyclization sequences. (h) The terminal stem-loop structure (3' SL) is found at the 3' end of all members of the genus *Flavivirus*.

ourselves to major structural entities encompassing the above-mentioned functional RNA elements, purposefully ignoring (small) stem-loop structures that are not conserved consistently in all MBFVs. Figure 5.3 shows example consensus structure predictions of these elements computed from structural alignments. We use viruses in the Japanese encephalitis virus group as examples here, as this group contains representatives of many structured RNAs found throughout the MBFV 3′ UTRs.

Stem-loops II and IV (Figure 5.3b and c) within domain I fold into three-way junction structures that exhibit exoribonuclease-stalling activity upon formation of a transient pseudoknot from their apical loop to downstream nucleotides (depicted in Figure 5.2). While both elements constitute central xrRNA building blocks, the SL-II apical loop is larger, thus facilitating more pseudoknot interactions among nucleotides than SL-IV. Importantly, structural homologs of SL-II/SL-IV are present in all MBFV species considered here, as well as in many other flaviviruses [23,39].

Another stable structural element in domain I is stem-loop III (SL-III, Figure 5.3d), which is conserved in all JEVG and many NTAVG viruses. Surprisingly, SL-III is also conserved in several dual-host affiliated insect-specific flaviviruses, including Nounané virus (NOUV), Barkedji virus (BJV), and Barkedji-like virus (BJLV) [32]. SL-III has a length between 66 and 76 nt and, within MBFVs, is always flanked by the xrRNA-associated SL-II and SL-IV elements. Although supported by rich covariation patterns (Figure 5.3e), the biological function of this thermodynamically stable element remains elusive.

Dumbbell (DB) elements (Figure 5.3f and g) are typically found in one or two copies and represent major building blocks of domain II in many MBFVs, TBFVs, and NKVs [23]. These well-conserved structures are characterized by high sequence variability in the proximal (more upstream) stem-loop originating from the central multi-loop, and, conversely, high primary sequence conservation in the distal (more downstream) stem-loop. This pattern is observed in all known DB elements [41] and rooted in the fact that the distal stem-loop overlaps cyclization sequences required by the viruses to form proper pan-handle structures during replication [42,56].

The terminal domain III of the 3′ UTR is characterized by a high degree of structural conservation throughout all ecological groups of flaviviruses. This region contains a small hairpin loop of ~15 nt length, which is immediately followed by a longer 3′ stem-loop structure (Figure 5.3h). We consider the two structural elements together here, designating the combined element as 3′ SL.

In the following Sections 5.4.1–5.4.7 we provide a qualitative description of RNA structure conservation within MBFV 3' UTRs, without focusing on known functional details. We show for each antigenic complex a plot comprised of a phylogenetic tree computed from complete coding sequence nucleotide alignments (sensu Figure 5.1) and a schematic of the 3' UTR organization for each virus species. Colored regions in these plots highlight evolutionarily conserved elements. Additionally, we provide 3' UTR predictions for some species to facilitate contextualization of elements that show pervasive conservation versus those that are only found in specific groups. Features of 3' UTR regions shown here refer to virus species listed in Table 5.1. No data are shown for the Edge hill virus group, as our data set does not include any 3' UTR sequences for this clade.

5.4.1 Japanese Encephalitis Virus Group

We begin the characterization of evolutionarily conserved RNAs in the 3' UTR of different MBFV complexes with the Japanese encephalitis virus group (Figure 5.4). The JEVG comprises serologically related virus species, encompassing medically important human pathogens, including JEV, WNV, USUV, St. Louis encephalitis virus (SLEV), and Murray Valley encephalitis virus (MVEV). The 3' UTR lengths in JEVG range from 549 to 665 nt, with SLEV having the shortest, and USUV having the longest 3' UTR, respectively. While the available 3' UTR sequences of Koutango virus (KOUV) and Cacipacore virus (CPCV) are truncated at the 3' end, variability in the JEVG 3' UTR lengths is largely due to variability in domain I, at the 5' end of the 3' UTR. No 3' UTR sequences are available for Yaoundé virus (YAOV).

Comparison of the JEVG 3' UTRs highlights a common architecture of functional RNA elements. All viruses in this group, with the exception of CPCV, contain stem-loops II, III, and IV, followed by duplicated DB elements and a terminal 3'-stem loop (3' SL) element in this order (Figure 5.4). Domain I in CPCV stands out, as it only contains SL-II and SL-III elements, but no SL-IV structure. Intriguingly, the 3' UTR region upstream of SL-II is unexpectedly long in CPCV, with pairwise distances between the annotated elements in good agreement with other JEVG viruses. Together with the missing SL-IV, this raises the question of whether CPCV has evolved different 3' UTR traits. Contrary, domain I in CPCV contains many potential stop codons, which could indicate that the effective NS5 stop codon is located more downstream.

Another interesting observation is the length bias among WNV lineages 1 (WNV-1) and 2 (WNV-2). While WNV-1 and the related Kunjin

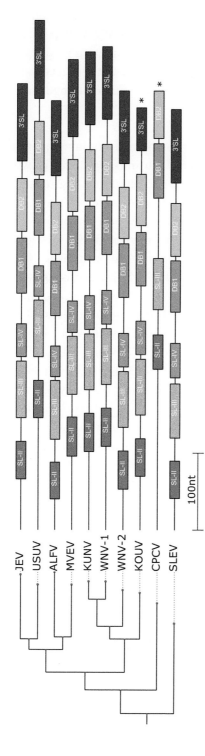

FIGURE 5.4 Annotated 3′ UTR of the Japanese encephalitis virus group. All species in this group except CPCV show a conserved architectural arrangement of functional RNAs following the pattern SL-II, SL-III, SL-IV, DB1, DB2, 3′ SL, with homogeneous spacer region lengths between the functional elements. Variable regions upstream of SL-II harbor structured RNAs in only a few species, e.g., SL-I in USUV (Figure 5.2), and are not shown here. Truncated sequences are marked by an asterisk *.

virus (KUNV) show an almost identical 3′ UTR architecture with a length of 631 and 624 nt, respectively; domain I of WNV-2 is almost 60 nt shorter. 3′ UTR architectural variability among different strains or lineages of a particular species is not unusual in RNA viruses and has been observed, e.g., in tick-borne encephalitis virus [57] and chikungunya virus [18].

5.4.2 Ntaya Virus Group

The Ntaya virus group comprises human and avian pathogenic viruses that are primarily maintained in transmission cycles between *Culex* spp. mosquitoes and birds. The entire NTAVG forms a sister clade of JEVG in the complete coding sequence nucleotide-based phylogeny (Figure 5.1) and internally splits into two subclades. One of them contains the type species, Ntaya virus (NTAV), as well as Bagaza virus (BAGV), Israel turkey meningoencephalomyelitis virus (ITV), and Tembusu virus (TMUV) with its subtypes Duck Tembusu virus (DTMUV), Sitiawan virus (SV), and Duck egg drop syndrome virus (DEDSV, not shown here); and the other contains Ilheus virus (ILHV) together with its subtype Rocio virus (ROCV), and T'Ho virus (THOV) (Figure 5.5). While 3′ UTR sequences are available for all of the above viruses except for SV, those of ITV and NTAV are 3′ truncated in our data set. 3′ UTR lengths range from 388 nt (ILHV) to 692 nt (DTMUV).

The clade that includes NTAV shows a relatively homogeneous architecture of structured elements in domain I of the 3′ UTR, encompassing SL-II, SL-III, and SL-IV elements. In domain II, BAGV and the truncated ITV form a single DB element, while the other viruses exhibit duplicated DB elements. In the clade comprising THOV, ILHV, and ROCV, no SL-IV elements are found. SL-III is only found in THOV, which moreover exhibits two copies of SL-II, both of which downstream of SL-III. A terminal 3′ SL is commonly present in all NTAVG species.

5.4.3 Aroa Virus Group

The Aroa virus group contains four recognized species [58], Aroa virus (AROAV), Iguape virus (IGUV), Bussuquara virus (BSQV, not shown here), and Naranjal virus (NJLV). 3′ UTR sequences are only available for two members, AROAV and IGUV, with a length of 421 and 567 nt, respectively (Figure 5.6). While the IGUV 3′ UTR contains both SL-II and SL-IV elements in domain I, which are separated by ~125 nt, AROAV contains only a single SL-II element. Domains II and III contain both DB elements and a canonical 3′ SL in both viruses.

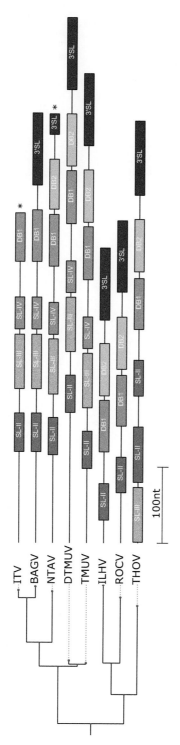

FIGURE 5.5 Annotated 3′ UTR of viruses in the Ntaya virus group.

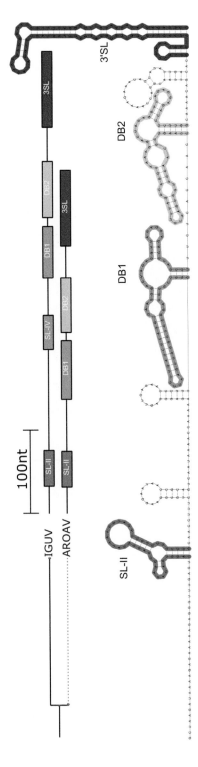

FIGURE 5.6 Top: 3′ UTR organization in the Aroa virus group. Bottom: Secondary structure prediction of the AROAV 3′ UTR. Conserved elements are highlighted in colors that match those in the top panel.

5.4.4 Kokobera Virus Group

The Kokobera virus group comprises five recognized species, which are all found in Australia and Papua New Guinea: Kokobera virus (KOKV), Bainyik virus (BAIV), Stratford virus (STRV), Torres virus (TORV), and New Mapoon virus (NMV). Only two of these (KOKV and NMV) have complete 3′ UTR sequences in the database, with a length of 558 and 546 nt, respectively. The truncated 3′ UTR sequences of the other species range from 233 nt (STRV) to 414 nt (TORV).

Homology search within KOKVG revealed a strictly homogeneous 3′ UTR organization (Figure 5.7), encompassing a unique pattern of structure conservation in domain I: The SL-II and SL-IV elements enclose another stem-loop element (designated SL-IIIk here), that is exclusively present in KOKVG and does not show structural homology to the SL-III elements found in JEVG and NTAVG. The SL-IIIk element is ~66nt long and folds into a bulged stem-loop structure (Figure 5.8). However, due to high sequence similarity in the 3′ UTR of the KOKVG viruses, there is only moderate covariation support for this element (data not shown). Nevertheless, the SL-IIIk element is thermodynamically stable and its presence in all known KOKVG members hints toward unknown biological function. The remaining 3′ UTR regions resemble the above discussed MBFV groups, specifically containing two DB elements and a terminal 3′ SL structure.

5.4.5 Dengue Virus Group

The Dengue virus group, which comprises four serotypes of Dengue virus (DENV1 – DENV4), is probably the best studied of all Flavivirus complexes, given that the pathology induced by the Dengue viruses represents the most prevalent arthropod-borne disease worldwide. Dengue viruses exhibit a highly homogeneous 3′ UTR architecture, with DENV1 being the longest (462 nt) and DENV4 being the shortest (384 nt) sequence. The Dengue 3′ UTRs typically contain SL-II and SL-IV and both DB elements, as well as the terminal 3′ SL element common to all flaviviruses (Figure 5.9).

The two SL copies in domain I of DENV1, DENV2, and DENV3 have been extensively studied and were used as model systems to elucidate xrRNA functionality [34]. Notably, DENV4 stands out in this region of the 3′ UTR, as it only contains a single SL element. Careful inspection revealed that the DENV4 SL exhibits higher similarity to SL-IV than

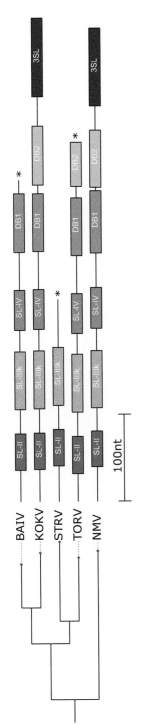

FIGURE 5.7   Annotated 3′ UTR of viruses in the Kokobera virus group.

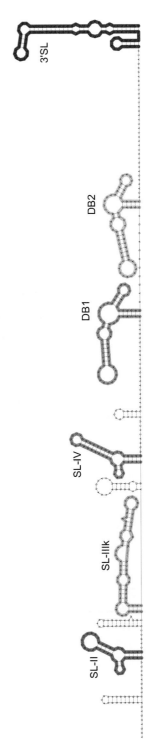

FIGURE 5.8   Secondary structure prediction of the KOKV 3′ UTR. Evolutionarily conserved elements are highlighted in colors that match those in Figure 5.7.

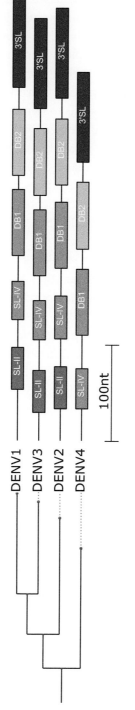

FIGURE 5.9   Annotated 3′ UTR of viruses in the Dengue virus group.

SL-II, as verified by covariance model-based comparison to SL elements in other flaviviruses. The presence of just a single xrRNA-associated SL-IV element in DENV4 is unexpected, as it has been suggested that two SL copies account for vector/host specificity and provide backup functionality to ensure proper sfRNA generation in case exoribonuclease halting is not achieved by the first SL copy [39]. Nevertheless, the presence of a single SL-IV copy in DENV4 allows us to hypothesize that this virus has lost the other SL element due to unknown evolutionary constraints. The remaining parts of DENVG 3′ UTRs are highly homogeneous, exhibiting duplicated DB elements and a terminal 3′ SL structure. The duplicated DB elements present in all four Dengue serotypes have been actively studied, and it was hypothesized that two DB elements are required in flaviviruses that have to replicate in different species [56].

5.4.6 Spondweni/Kedougou Virus Group

Another antigenic flavivirus complex that has attracted considerable research interest in recent years is the Spondweni virus group, containing Spondweni virus (SPOV) and Zika virus (ZIKV). The latter is considered an emerging pathogen, being responsible for a large outbreak in the Americas 2015–2017 that resulted in more than 1.5 million infections in Brazil alone and ~3,500 cases of congenital microcephaly [59]. SPOVG shares ancestral roots with DENVG (Figure 5.1), and although it formally contains only two species, we consider it here together with Kedougou virus (KEDV), which phylogenetically clusters with SPOV and ZIKV. We examine the 3′ UTRs of two ZIKV lineages: the ancestral African lineage based on an isolate from Uganda (ZIKV-UG, 428 nt) and the American lineage based on a Brazilian isolate (ZIKV-BR, 429 nt) (Figure 5.10). The 3′ UTR of SPOV is 338 nt long and truncated within the 3′ SL element. KEDV has a 3′ UTR of 390 nt length.

Comparison of the 3′ UTRs of SPOVG reveals a variable 3′ UTR organization in domains I and II, and a common 3′ SL element in domain III. The ZIKV 3′ UTR architecture is particularly interesting, as it stands out among MBFVs. While ZIKV has duplicated xrRNA-associated SL elements in domain I, and bona fide exoribonuclease stalling capacity has been experimentally confirmed [60], our covariance model screen revealed that the second SL element is fairly related to DENVG SL-II, while the first SL (Z.SL-I) is only remotely related to other SL elements. This suggests that Z.SL-I emerged from a ZIKV-specific duplication or recombination event. Conversely, domain I in SPOV and KEDV contain a single SL-II

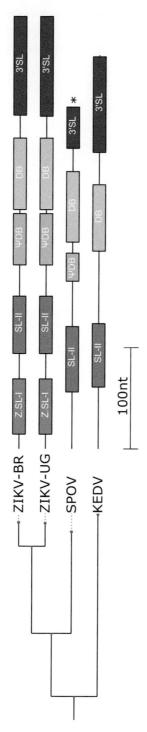

FIGURE 5.10 Annotated 3′ UTR of viruses in the Spondweni virus group.

structure each, without evidence for the presence of a ZIKV-like SL-I element. Domain II of the SPOVG viruses is no less interesting, as it contains only a single canonical DB element preceded by a pseudo-DB (ΨDB) element in ZIKV and SPOV, but not in KEDV. Importantly, the ΨDB element not homologous to DB elements found in ZIKV or any other MBFV species (see Section 5.4.8). The terminal 3′ SL structure is again found in all SPOVG viruses.

5.4.7 Yellow Fever Virus Group

Two antigenic complexes within the MBFVs, the Yellow fever virus group and the Edge hill virus group are located ancestral to the other MBFV groups in the complete coding sequence nucleotide phylogeny (Figure 5.1). The YFVG includes the species Yellow fever virus (YFV), Wesselsbron virus (WESSV), Fitzroy River virus (FRV), and Sepik virus (SEPV), with 3′ UTR lengths ranging from 459 nt (SEPV) to 657 nt (YFV). Several genotypes of YFV are known that differ in replicative fitness and virulence [61], and whose 3′ UTRs differ in length and exhibit structural heterogeneity [62,63]. We consider here an isolate of the YFV South American I genotype as well as the live attenuated YFV 17D strain to showcase within-species 3′ UTR architectural diversity.

The YFVG 3′ UTR organization is in qualitative agreement with that of other MBFV groups, however, there are several characteristics that make YFVG 3′ UTRs unique. Probably the most obvious is the presence of a variable number of imperfect sequence repeats (RYFs) of approximately 40nt length that fold into simple stem-loop strictures at the beginning of the 3′ UTR (Figure 5.11). While these sequence repeats have been predicted to fold into two small, consecutive hairpins (denoted G and F) in the literature [64], our predictions suggest that only the downstream part (F hairpin) forms a thermodynamically stable structure in a genomic context. The upstream repeat region (G), however, is involved in the formation of specific folds that are not structurally conserved throughout the YFVG. Focusing on structure conservation, we only consider the downstream (F) hairpin of the RYF elements with a length between 20 and 22 nt. RYF elements occur in variable copy numbers, depending on species and genotype, and show a good covariation support. Figure 5.12 shows the consensus secondary structure of the RYF hairpin and illustrates the formation of three RYF hairpins in the 3′ UTR of Fitzroy River virus. RYF upstream regions (G region) of these elements do not fold into conserved structures.

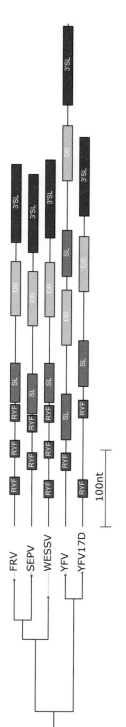

FIGURE 5.11   Annotated 3′ UTR of viruses in the Yellow fever virus group.

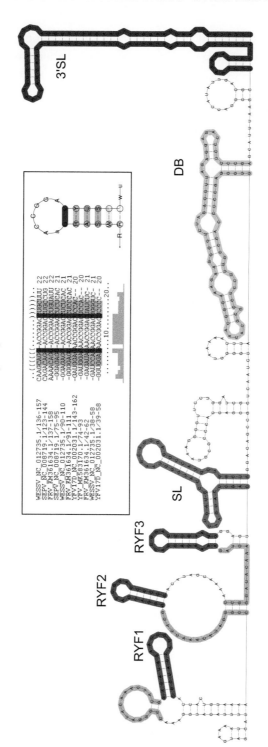

FIGURE 5.12   Secondary structure prediction of the Fitzroy River virus 3′ UTR. Colored structurally conserved elements match those depicted in Figure 5.11. Structurally non-conserved G repeats are shown in faint red upstream of the RYF hairpins. The insert shows a structural alignment and consensus structure prediction of 3′ half of RYF elements (F hairpin). Coloring of the insert follows the **RNAalifold** scheme, highlighting different covariation levels (Figure 5.3a).

Downstream of the RYF-containing region, an xrRNA-associated three-way-junction forming stem-loop element (denoted SL-E in literature) is found in many YFVG species. This element confers the protection of downstream regions against exoribonuclease degradation [65] and exhibits a high similarity to the NTAVG SL-II structure. Moreover, a DB element that remotely resembles the DB elements of the other MBFV antigenic groups, but with a longer proximal arm, is found in all members of the YFVG groups, as well as the obligatory terminal 3′ SL structure. Notably, the region that contains the SL-E and DB elements is duplicated in the YFV isolate discussed here (Figure 5.11).

While YFVG SL-E and DB elements appear to exhibit similar biological functionality to their remote homologs in the other flaviviruses, their structural divergence is likely a result of their evolutionary distance to the other MBFV groups. This is also suggested by the positioning of the YFVG clade in the phylogenetic tree relative to the other MBFV groups.

5.4.8 Structural Diversity of Conserved Elements

Structural similarity among duplicated elements has previously been addressed by tree alignment approaches. In the case of DENV, it has been suggested that orthologous DB elements share a higher similarity than paralogous DB elements. Specifically, DB1 elements (and DB2 elements, respectively) of different Dengue serotypes are more alike among themselves than DB1 and DB2 elements from the same serotype [56]. Similar examples, encompassing xrRNA-associated SL elements have been proposed in the four Dengue serotypes and several species of the JEVG [39], suggesting divergent evolutionary trajectories after the duplication events.

We revisit this topic here and present a comprehensive comparison of functional RNA elements across all antigenic MBFV groups, focusing specifically on DB elements and xrRNA-associated SL elements. Rather than comparing the sequences and structures of individual strains, we directly compare the covariance models built from alignments of representative sequences in each antigenic group. We use **CMCompare** [54], which for two given covariance models finds the link sequence, defined as the sequence giving the highest score on both models simultaneously. The associated link score is a measure for the similarity of the models that considers both sequence and structure similarity of the two groups. Models for closely related elements will exhibit high link scores, while low or even negative link scores indicate unrelatedness of the models.

The link scores computed from all pairwise comparisons of DB element covariance models discussed in this contribution are shown in Figure 5.13 as a similarity matrix. Structural proximity among DB elements as well as the SPOVG psiDB element is color-coded, highlighting a dense network of structural similarity among duplicated DB elements. Intriguingly, our data do not support the hypothesis that orthologous entities are generally more alike than paralogous entities. Rather, most DB elements exhibit high link scores to either orthologous or paralogous elements, suggesting that they could also be captured by a single covariance model.

Comparison of the SPOVG DB element with all other DB elements reveals slightly higher link scores to DB2 elements than DB1 elements. Likewise, the DB element present in the YFVG has only very weak similarity to the other DB elements, with the highest link score to the DB2 of KOKVG. Importantly, the SPOVG psiDB element has only negative or very low link scores <2, indicating that this element is not homologous to any other DB element.

The situation is different for the comparison of xrRNA-associated SL elements (denoted SL elements here). As shown in Figure 5.14, our data do not suggest pervasive similarity among the majority of elements, but rather a partitioning into distinct sets that exhibit high intra-set scores, i.e., link scores to elements of the same set. Conversely, the link scores between SL elements of different sets, i.e., inter-set scores, are consistently low. The set with the highest intra-set scores comprises the SL-II and SL-IV elements of JEVG, as well as the SL-II of NTAVG, and the single YFVG SL element. Interestingly, the NTAVG SL-IV is not part of this set but has only weak link scores to the other SL elements. The second group of elements with high intra-set link-scores encompasses DENVG, KOKVG, and AROAVG, each with their SL-II and SL-IV elements.

While we could not identify an example where the highest link score of an element is with its tandem copy in the same species, there are cases where we see fairly high link scores among paralogous elements, e.g., the SL-II/SL-IV elements of JEVG and KOKVG. Importantly, the observation that several elements have their highest link score with orthologous elements of viruses within the same set/cluster, and at the same time, these elements have particularly low link scores with their orthologs from the other cluster argues against the hypothesis that a fixed architecture of duplicated SL has been conserved from the last common ancestor of all MBFVs. Rather, these data suggest a scenario of repeated duplication and loss events of SL elements, which is also supported by the weak similarity

	SPOVG.psiDB	YFVG.DB	SPOVG.DB	NTAVG.DB2	JEVG.DB2	KOKVG.DB2	DENVG.DB2	AROAVG.DB2	NTAVG.DB1	JEVG.DB1	KOKVG.DB1	DENVG.DB1	AROAVG.DB1
SPOVG.psiDB	63.11	-1.61	0.44	-0.02	-0.17	0.12	0.05	-0.66	-0.66	-0.66	-0.21	-0.66	-0.66
YFVG.DB	-1.61	99.08	17.59	13.99	10.93	27.59	12.67	16.74	1.39	5.19	10.24	23.80	2.82
SPOVG.DB	0.44	17.59	102.22	70.59	60.90	64.84	64.12	65.50	60.37	57.80	57.79	59.17	46.09
NTAVG.DB2	-0.02	13.99	70.59	100.33	72.45	76.69	86.71	76.67	71.51	74.79	70.79	56.84	51.61
JEVG.DB2	-0.17	10.93	60.90	72.45	97.15	62.16	69.32	64.67	59.43	59.17	56.10	62.88	65.56
KOKVG.DB2	0.12	27.59	64.84	76.69	62.16	99.84	77.92	80.06	59.43	65.07	74.44	60.39	38.34
DENVG.DB2	0.05	12.67	64.12	86.71	69.32	77.92	97.48	77.43	62.94	70.42	66.53	55.59	45.93
AROAVG.DB2	-0.66	16.74	65.50	76.67	64.67	80.06	77.43	99.39	54.49	58.73	70.20	59.70	36.95
NTAVG.DB1	-0.66	1.39	60.37	71.51	59.43	59.43	62.94	54.49	96.81	71.13	63.93	46.59	53.11
JEVG.DB1	-0.66	5.19	57.80	74.79	59.17	65.07	70.42	58.73	71.13	97.74	75.33	51.08	48.46
KOKVG.DB1	-0.21	10.24	57.79	70.79	56.10	74.44	66.53	70.20	63.93	75.33	96.60	47.54	44.25
DENVG.DB1	-0.66	23.80	59.17	56.84	62.88	60.39	55.59	59.70	46.59	51.08	47.54	101.52	40.92
AROAVG.DB1	-0.66	2.82	46.09	51.61	65.56	38.34	45.93	36.95	53.11	48.46	44.25	40.92	91.12

score 0   25   50   75   100

FIGURE 5.13   Similarity matrix of DB elements computed from pairwise covariance model comparisons. Link score values highlighted as different shades of olive indicate the similarity levels between pairs. Diagonal entries represent the score that can be achieved by comparing a model to itself, i.e. the score of the best link sequence.

	SPOVG.SL-II	ZIKV.SL-I	NTAVG.SL-IV	JEVG.SL-IV	NTAVG.SL-II	JEVG.SL-II	YFVG.SL	AROAVG.SL-II	KOKVG.SL-II	DENVG.SL-II	KOKVG.SL-IV	AROAVG.SL-IV	DENVG.SL-IV
SPOVG.SL-II	88.82	3.72	-5.08	1.97	0.80	-0.89	1.72	-2.89	-6.13	16.94	2.55	-0.04	-3.15
ZIKV.SL-I	3.72	88.08	-2.62	13.22	10.46	12.31	12.21	-2.24	-2.62	-1.64	8.80	-2.62	5.96
NTAVG.SL-IV	-5.08	-2.62	72.58	10.76	9.84	9.84	-4.19	7.78	7.78	7.82	4.04	6.89	7.78
JEVG.SL-IV	1.97	13.22	10.76	70.11	42.34	37.00	26.40	8.42	8.70	8.70	8.06	6.89	8.70
NTAVG.SL-II	0.80	10.46	9.84	42.34	78.49	60.23	39.23	8.42	8.61	8.61	6.59	8.25	8.61
JEVG.SL-II	-0.89	12.31	9.84	37.00	60.23	78.41	26.47	8.42	9.42	9.37	6.06	6.89	9.42
YFVG.SL	1.72	12.21	-4.19	26.40	39.23	26.47	80.49	-1.70	6.49	1.31	4.62	1.25	1.57
AROAVG.SL-II	-2.89	-2.24	7.78	8.42	8.42	8.42	-1.70	66.12	26.97	16.25	18.22	16.69	10.03
KOKVG.SL-II	-6.13	-2.62	7.78	8.70	8.61	9.42	6.49	26.97	73.06	41.77	34.73	31.94	16.22
DENVG.SL-II	16.94	-1.64	7.82	8.70	8.61	9.37	1.31	16.25	41.77	72.89	36.77	33.31	23.50
KOKVG.SL-IV	2.55	8.80	4.04	8.06	6.59	6.06	4.62	18.22	34.73	36.77	71.04	24.87	26.81
AROAVG.SL-IV	-0.04	-2.62	6.89	6.89	8.25	6.89	1.25	16.69	31.94	33.31	24.87	59.81	13.74
DENVG.SL-IV	-3.15	5.96	7.78	8.70	8.61	9.42	1.57	10.03	16.22	23.50	26.81	13.74	70.68

score 0 20 40 60 80

FIGURE 5.14 Link scores computed from pairwise covariance model comparisons of xrRNA-associated SL elements. Diagonal entries show the maximal score that can be achieved when comparing a motel with itself. See text for details.

of SPOVG SL elements: While the SPOVG SL-II exhibits the highest similarity to the DENVG SL-II element, our data show that the ZIKV SL-I structure is only very distantly related to the other SL elements, suggesting that this could have emerged from a ZIKV-specific duplication or recombination event.

5.5 CONCLUSION

Flavivirus 3′ UTRs have increasingly attracted research interest over the last years, both experimentally and theoretically. These regions are special, as they exhibit a striking level of plasticity that is manifested in a varied pattern of conserved RNA elements whose functional associations have been solved for some flaviviruses. A profound knowledge about the structural landscape of these genomic regions is crucial to understanding the mechanistic traits that modulate the viral life cycle and pathology.

The availability of large numbers of viral genome data over the last years has made available the possibility to perform comparative genomics screens at an unprecedented level of detail, allowing researchers to elucidate hitherto unknown structural traits in silico. In this context, comparative genomics screens based on structural alignments and RNA family models turned out as a particularly useful approach for the detection of homologous structures, thereby overcoming the limitations of sequence-based homology search approaches. Recently, an alignment-free, descriptor-based method for structured RNA pattern search has been proposed [66]. Such methods allow to include tertiary interactions like pseudoknots or base triplets that cannot be covered in covariance models. In contrast to covariance models which are automatically built from a consensus structure and sequence alignment, descriptor models have to be handcrafted. Thus both methods have their specific advantages and can complement each other.

Many RNA viruses have successfully evolved strategies to tolerate error-prone replication, resulting in increased mutation frequencies that allow them to evade host immune responses. Likewise, there is evolutionary pressure on maintaining functional entities, such as structured elements in the UTRs that are crucially involved in virus replication and mediating pathogenesis. Knowledge about the exact location of these evolutionarily conserved elements within the 3′ UTR of each antigenic group allows not only to characterize homologous structures in novel strains but also to infer the proximity between functionally conserved RNAs.

The presence of multiple copies of functional RNA elements in many flavivirus 3′ UTRs likely resulted from repeated duplication and loss events, yielding diversified structures that allow the virus to adapt to its specific combination of host and vector. The presence of tandem copies (or even higher copy numbers in some insect-specific flaviviruses [23]) of SL and DB elements raises the question of why a lot of viruses maintain potentially redundant parts of their genomes in spite of the strong seletion pressure on genome length. Some evidence suggests that duplicated elements may serve specific roles in different stages of the viral life cycle, such as vector and host stages, [40,56,67]. Knowledge about a particular instance of an element in a specific virus can be extrapolated to related species. The obvious quest is to derive functional knowledge from studied examples and extrapolate this to newly discovered or (re-)emerging viruses. This is particularly useful for addressing open questions like the functional association of structured RNAs with vector/host specificity. Moreover, a detailed understanding of the proximity among common structural patterns in different viruses is crucial for studying virus UTR evolution.

REFERENCES

1. Scott C Weaver, Caroline Charlier, Nikos Vasilakis, and Marc Lecuit. Zika, chikungunya, and other emerging vector-borne viral diseases. *Annu. Rev. Med.*, 69:395–408, 2018.
2. Bradley Blitvich and Andrew Firth. Insect-specific flaviviruses: A systematic review of their discovery, host range, mode of transmission, superinfection exclusion potential and genomic organization. *Viruses*, 7(4):1927–1959, 2015.
3. Peter M Howley and David M Knipe, editors. *Fields Virology: Emerging Viruses*. Lippincott Williams & Wilkins: Philadelphia, PA, 7th edition, 2020.
4. Bradley Blitvich and Andrew Firth. A review of flaviviruses that have no known arthropod vector. *Viruses*, 9(6):154, 2017.
5. Rhys Parry and Sassan Asgari. Discovery of novel crustacean and cephalopod flaviviruses: Insights into the evolution and circulation of flaviviruses between marine invertebrate and vertebrate hosts. *J. Virol.*, 93(14):e00432–19, 2019.
6. Theodore C Pierson and Michael S Diamond. The continued threat of emerging flaviviruses. *Nat. Microbiol.*, 5(6):796–812, 2020.
7. Franz X Heinz and Karin Stiasny. Flaviviruses and flavivirus vaccines. *Vaccine*, 30(29):4301–4306, 2012.
8. Gilda Grard, Gregory Moureau, Remi N Charrel, Edward C Holmes, Ernest A Gould, and Xavier de Lamballerie. Genomics and evolution of Aedes-borne flaviviruses. *J. Gen.Virol.*, 91(1):87–94, 2010.
9. Wy Ng, Ruben Soto-Acosta, Shelton Bradrick, Mariano Garcia-Blanco, and Eng Ooi. The 5′ and 3′ untranslated regions of the flaviviral genome. *Viruses*, 9(6):137, 2017.

10. Charles M Rice, Edith M Lenches, Sean R Eddy, Sung Joon Shin,Raymond L Sheets, and James H Strauss. Nucleotide sequence of yellow fever virus: Implications for flavivirus gene expression and evolution. *Science,* 229(4715): 726–733, 1985.

11. Sergio M Villordo, Diego E Alvarez, and Andrea V Gamarnik. A balance between circular and linear forms of the dengue virus genome is crucial for viral replication. *RNA,* 16(12):2325–2335, 2010.

12. Luana de Borba, Sergio M Villordo, Nestor G Iglesias, Claudia V Filomatori, Leopoldo G Gebhard, and Andrea V Gamarnik. Overlapping local and long-range RNA-RNA interactions modulate dengue virus genome cyclization and replication. *J. Virol.,* 89(6):3430–3437, 2015.

13. Chang S Hahn, Young S Hahn, Charles M Rice, Eva Lee, Lynn Dalgarno, Ellen G Strauss, and James H Strauss. Conserved elements in the 3′ untranslated region of flavivirus RNAs and potential cyclization sequences. *J. Mol. Biol.,* 198(1):33–41, 1987.

14. Margo A Brinton and Mausumi Basu. Functions of the 3′ and 5′ genome RNA regions of members of the genus flavivirus. *Virus Res.,* 206:108–119, 2015.

15. Michael Kiening, Roman Ochsenreiter, Hans-Jörg Hellinger, Thomas Rattei, Ivo Hofacker, and Dmitrij Frishman. Conserved secondary structures in viral mRNAs. *Viruses,* 11(5):401, 2019.

16. Christina Witwer, Susanne Rauscher, Ivo L Hofacker, and Peter F Stadler. Conserved RNA secondary structures in picornaviridae genomes. *Nucleic Acids Res.,* 29(24):5079–5089, 2001.

17. Caroline Thurner, Christina Witwer, Ivo L Hofacker, and Peter F Stadler. Conserved RNA secondary structures in flaviviridae genomes. *J. Gen. Virol.,* 85(5):1113–1124, 2004.

18. Adriano de Bernardi Schneider, Roman Ochsenreiter, Reilly Hostager, Ivo L. Hofacker, Daniel Janies, and Michael T. Wolfinger. Updated phylogeny of chikungunya virus suggests lineage-specific RNA architecture. *Viruses,* 11:798, 2019.

19. Dong Yang and Julian L Leibowitz. The structure and functions of coronavirus genomic 3′ and 5′ ends. *Virus Res.,* 206:120–133, 2015.

20. Ramakanth Madhugiri, Markus Fricke, Manja Marz, and John Ziebuhr. RNA structure analysis of alphacoronavirus terminal genome regions. *Virus Res.,* 194:76–89, 2014.

21. Ramakanth Madhugiri, Nadja Karl, Daniel Petersen, Kevin Lamkiewicz, Markus Fricke, Ulrike Wend, Robina Scheuer, Manja Marz, and John Ziebuhr. Structural and functional conservation of cis-acting RNA elements in coronavirus 5′-terminal genome regions. *Virology,* 517:44–55, 2018.

22. Alexandra Popa, Jakob-Wendelin Genger, Michael D. Nicholson, Thomas Penz, Daniela Schmid, Stephan W Aberle, Benedikt Agerer, Alexander Lercher, Lukas Endler, Henrique Colaco, et al. Genomic epidemiology of superspreading events in Austria reveals mutational dynamics and transmission properties of SARS-CoV-2. *Sci. Transl. Med.,* 12 (573):2555, 2020.

23. Roman Ochsenreiter, Ivo L Hofacker, and Michael T Wolfinger. Functional RNA structures in the 3′ UTR of tick-borne, insect-specific and no-known-vector flaviviruses. *Viruses*, 11(3):298, 2019.

24. Gorben P Pijlman, Anneke Funk, Natasha Kondratieva, Jason Leung, Shessy Torres, Lieke Van der Aa, Wen Jun Liu, Ann C Palmenberg, Pei-Yong Shi, Roy A Hall, et al. A highly structured, nuclease-resistant, noncoding RNA produced by flaviviruses is required for pathogenicity. *Cell Host Microbe*, 4(6):579–591, 2008.

25. Benjamin M Akiyama, Daniel Eiler, and Jeffrey S Kieft. Structured RNAs that evade or confound exonucleases: Function follows form. *Curr. Opin. Struc. Biol.*, 36:40–47, 2016.

26. Andrii Slonchak and Alexander A Khromykh. Subgenomic flaviviral RNAs: What do we know after the first decade of research. *Antivir. Res.*, 159:13–25, 2018.

27. Christopher Iain Jones, Maria Vasilyevna Zabolotskaya, and Sarah Faith Newbury. The 5′ → 3′ exoribonuclease Xrn1/pacman and its functions in cellular processes and development. *RNA*, 3(4):455–468, 2012.

28. Brian D Clarke, Justin A Roby, Andrii Slonchak, and Alexander A Khromykh. Functional non-coding RNAs derived from the flavivirus 3′ untranslated region. *Virus Res.*, 206:53–61, 2015.

29. Claudia V Filomatori, Juan M Carballeda, Sergio M Villordo, Sebastian Aguirre, Horacio M Pallarés, Ana M Maestre, Irma Sánchez-Vargas, Carol D Blair, Cintia Fabri, Maria A Morales, Ana Fernandez-Sesma, and Andrea V. Gamarnik. Dengue virus genomic variation associated with mosquito adaptation defines the pattern of viral non-coding RNAs and fitness in human cells. *PLoS Pathog.*, 13(3):e1006265, 2017.

30. Andrea MacFadden, Zoe O'Donoghue, Patricia A G C Silva, Erich G Chapman, René C Olsthoorn, Mark G Sterken, Gorben P Pijlman, Peter J Bredenbeek, and Jeffrey S Kieft. Mechanism and structural diversity of exoribonuclease-resistant RNA structures in flaviviral RNAs. *Nat. Commun.*, 9(1):119, 2018.

31. Rachel A. Jones, Anna-Lena Steckelberg, Matthew J. Szucs, Benjamin M. Akiyama, Quentin Vicens, and Jeffrey S. Kieft. Different tertiary interactions create the same important 3-D features in a divergent flavivirus xrRNA. *BioRxiv*, 2020.

32. Christida E. Wastika, Hayato Harima, Sasaki Michihito, Bernard M. Hang'ombe, Yuki Eshita, Qiu Yongjin, William W. Hall, Michael T. Wolfinger, Hirofumi Sawa, and Yasuko Orba. Discoveries of exoribonuclease-resistant structures of insect-specific flaviviruses isolated in Zambia. *Viruses*, 12:1017, 2020.

33. Xiaolin Niu, Qiuhan Liu, Zhonghe Xu, Zhifeng Chen, Linghui Xu, Lilei Xu, Jinghong Li, and Xianyang Fang. Molecular mechanisms underlying the extreme mechanical anisotropy of the flaviviral exoribonuclease-resistant RNAs (xrRNAs). *Nat. Commun.*, 11:5496, 2020.

34. Erich G Chapman, Stephanie L Moon, Jeffrey Wilusz, and Jeffrey S Kieft. RNA structures that resist degradation by Xrn1 produce a pathogenic Dengue virus RNA. *Elife*, 3:e01892, 2014.

35. Anneke Funk, Katherine Truong, Tomoko Nagasaki, Shessy Torres, Nadia Floden, Ezequiel Balmori Melian, Judy Edmonds, Hongping Dong, Pei-Yong Shi, and Alexander A Khromykh. RNA structures required for production of subgenomic flavivirus RNA. *J. Virol.*, 84(21):11407–11417, 2010.

36. Susanne Rauscher, Christoph Flamm, Christian W Mandl, Franz X Heinz, and Peter F Stadler. Secondary structure of the 3′-noncoding region of flavivirus genomes: Comparative analysis of base pairing probabilities. *RNA*, 3(7):779–791, 1997.

37. Ivo L. Hofacker, Martin Fekete, Christoph Flamm, Martijn A. Huynen, Susanne Rauscher, Paul E. Stolorz, and Peter F. Stadler. Automatic detection of conserved RNA structure elements in complete RNA virus genomes. *Nucleic Acids Res.*, 26(16):3825–3836, 1998.

38. Ivo L Hofacker, Peter F Stadler, and Roman R Stocsits. Conserved RNA secondary structures in viral genomes: A survey. *Bioinformatics*, 20(10):1495–1499, 2004.

39. Sergio M Villordo, Juan M Carballeda, Claudia V Filomatori, and Andrea V Gamarnik. RNA structure duplications and flavivirus host adaptation. *Trends Microbiol.*, 24(4):270–283, 2016.

40. Joanna Sztuba-Solinska, Tadahisa Teramoto, Jason W Rausch, Bruce A Shapiro, Radhakrishnan Padmanabhan, and Stuart F J Le Grice. Structural complexity of dengue virus untranslated regions: cis-acting RNA motifs and pseudoknot interactions modulating functionality of the viral genome. *Nucleic Acids Res.*, 203,5075–89, 2013.

41. Adriano de Bernardi Schneider and Michael T. Wolfinger. Musashi binding elements in Zika and related Flavivirus 3′ UTRs: A comparative study *in silico*. *Sci. Rep.*, 9(1):6911, 2019.

42. Alexander A Khromykh, Hedije Meka, Kimberley J Guyatt, and Edwin G Westaway. Essential role of cyclization sequences in flavivirus RNA replication. *J. Virol.*, 75(14):6719–6728, 2001.

43. Diego E Alvarez, María F Lodeiro, Silvio J Luduena, Lía I Pietrasanta, and Andrea V Gamarnik. Long-range RNA-RNA interactions circularize the dengue virus genome. *J. Virol.*, 79(11):6631–6643, 2005.

44. Claudia V Filomatori, Maria F Lodeiro, Diego E Alvarez, Marcelo M Samsa, Lía Pietrasanta, and Andrea V Gamarnik. A 5′ RNA element promotes dengue virus RNA synthesis on a circular genome. *Gene Dev.*, 20(16):2238–2249, 2006.

45. Susann Friedrich, Susanne Engelmann, Tobias Schmidt, Grit Szczepankiewicz, Sandra Bergs, Uwe G Liebert, Beate M Kümmerer, Ralph P Golbik, and Sven-Erik Behrens. The host factor AUF1 p45 supports flavivirus propagation by triggering the RNA switch required for viral genome cyclization. *J. Virol.*, 92(6):e01647–17, 2018.

46. Kazutaka Katoh and Daron M Standley. MAFFT multiple sequence alignment software version 7: Improvements in performance and usability. *Mol. Biol. Evol.*, 30(4):772–780, 2013.

47. Lam-Tung Nguyen, Heiko A Schmidt, Arndt von Haeseler, and Bui Quang Minh. IQ-TREE: A fast and effective stochastic algorithm for estimating maximum-likelihood phylogenies. *Mol. Biol. Evol.*, 32(1):268–274, 2014.

48. Eric P. Nawrocki and Sean R. Eddy. Infernal 1.1: 100-fold faster RNA homology searches. *Bioinformatics*, 29:2933–2935, 2013.

49. Sean R Eddy and Richard Durbin. RNA sequence analysis using covariance models. *Nucleic Acids Res.*, 22(11):2079–2088, 1994.

50. Sebastian Will, Kristin Reiche, Ivo L Hofacker, Peter F Stadler, and Rolf Backofen. Inferring noncoding RNA families and classes by means of genome-scale structure-based clustering. *PLoS Comp. Biol.*, 3(4):e65, 2007.

51. Sam Griffiths-Jones, Alex Bateman, Mhairi Marshall, Ajay Khanna, and Sean R Eddy. Rfam: An RNA family database. *Nucleic Acids Res.*, 31(1):439–441, 2003.

52. Stephan H Bernhart, Ivo L Hofacker, Sebastian Will, Andreas R Gruber, and Peter F Stadler. RNAalifold: Improved consensus structure prediction for RNA alignments. *BMC Bioinf.*, 9(1):474, 2008.

53. Ronny Lorenz, Stephan H Bernhart, Christian Hoener Zu Siederdissen, Hakim Tafer, Christoph Flamm, Peter F Stadler, and Ivo L Hofacker. Viennarna package 2.0. *Algorithm Mol. Biol.*, 6(1):26, 2011.

54. Christian Höner zu Siederdissen and Ivo L Hofacker. Discriminatory power of RNA family models. Bioinformatics, 26(18):i453–i459, 2010.

55. Clément Mazeaud, Wesley Freppel, and Laurent Chatel-Chaix. The multiples fates of the flavivirus RNA genome during pathogenesis. *Front. Genet.*, 9:595, 2018.

56. Luana de Borba, Sergio M Villordo, Franco L Marsico, Juan M Carballeda, Claudia V Filomatori, Leopoldo G Gebhard, Horacio M Pallarés, Sebastian Lequime, Louis Lambrechts, Irma Sánchez Vargas, Carol D. Blair, and Andrea V. Gamarnik. RNA structure duplication in the dengue virus 3′ UTR: Redundancy or host specificity? *mBio* 10(1):e02506–18, 2019.

57. Vladimir A Ternovoi, Anastasia V Gladysheva, Eugenia P Ponomareva, Tamara P Mikryukova, Elena V Protopopova, Alexander N Shvalov, Svetlana N Konovalova, Eugene V Chausov, and Valery B Loktev. Variability in the 3′ untranslated regions of the genomes of the different tick-borne encephalitis virus subtypes. *Virus Genes*, 55(4):448–457, 2019.

58. Gregory Moureau, Shelley Cook, Philippe Lemey, Antoine Nougairede, Naomi L Forrester, Maxim Khasnatinov, Remi N Charrel, Andrew E Firth, Ernest A Gould, and Xavier De Lamballerie. New insights into flavivirus evolution, taxonomy and biogeographic history, extended by analysis of canonical and alternative coding sequences. *PLoS One*, 10(2):e0117849, 2015.

59. Thalia Velho Barreto de Araújo, Ricardo Arraes de Alencar Ximenes, Demócrito de Barros Miranda-Filho, Wayner Vieira Souza, Ulisses Ramos Montarroyos, Ana Paula Lopes de Melo, Sandra Valongueiro, Cynthia Braga, Sinval Pinto Brandão Filho, Marli Tenório Cordeiro, et al. Association between microcephaly, Zika virus infection, and other risk factors in Brazil: Final report of a case-control study. *Lancet Infect Dis.*, 18(3):328–336, 2018.

60. Benjamin M Akiyama, Hannah M Laurence, Aaron R Massey, DavidKamit A Costantino, Xuping Xie, Yujiao Yang, Pei-Yong Shi, Jay C Nix, J David Beckham, and Jeffrey S Kieft. Zika virus produces noncoding RNAs using a multi-pseudoknot structure that confounds a cellular exonuclease. *Science*, 354:1148–1152, 2016.

61. Alan D T Barrett and Ernest A Gould. Comparison of neurovirulence of different strains of yellow fever virus in mice. *J. Gen. Virol.*, 67(4):631–637, 1986.

62. Juliet E Bryant, Pedro F C Vasconcelos, Rene C A Rijnbrand, John-Paul Mutebi, Stephen Higgs, and Alan D T Barrett. Size heterogeneity in the 3′ noncoding region of South American isolates of yellow fever virus. *J. Virol.*, 79(6):3807–3821, 2005.

63. Raphaëlle Klitting, Carlo Fischer, Jan Drexler, Ernest Gould, David Roiz, Christophe Paupy, and Xavier de Lamballerie. What does the future hold for yellow fever virus? (II). *Genes*, 9(9):425, 2018.

64. John-Paul Mutebi, René C A Rijnbrand, Heiman Wang, Kate D Ryman, Eryu Wang, Lynda D Fulop, Rick Titball, and Alan D T Barrett. Genetic relationships and evolution of genotypes of yellow fever virus and other members of the yellow fever virus group within the flavivirus genus based on the 3′ noncoding region. *J. Virol.*, 78(18):9652–9665, 2004.

65. Patrícia A G C Silva, Carina F Pereira, Tim J Dalebout, Willy J M Spaan, and Peter J Bredenbeek. An RNA pseudoknot is required for production of yellow fever virus subgenomic RNA by the host nuclease Xrn1. *J. Virol.*, 84(21):11395–11406, 2010.

66. Alan Zammit, Leon Helwerda, René C L Olsthoorn, Fons J Verbeek, and Alexander P Gultyaev. A database of flavivirus RNA structures with a search algorithm for pseudoknots and triple base interactions. *Bioinformatics*:1–7, 2020. https://doi.org/10.1093/bioinformatics/btaa759

67. Mark Manzano, Erin D Reichert, Stephanie Polo, Barry Falgout, Wojciech Kasprzak, Bruce A Shapiro, and Radhakrishnan Padmanabhan. Identification of cis-acting elements in the 3′-untranslated region of the dengue virus type 2 RNA that modulate translation and replication. *J. Biol. Chem.*, 286:22521–22534, 2011.

CHAPTER 6

Structural Bioinformatics of Influenza Virus RNA Genomes

Alexander P. Gultyaev

Leiden University
Erasmus Medical Center

René C.L. Olsthoorn

Leiden University

Monique I. Spronken and Mathilde Richard

Erasmus Medical Center

CONTENTS

6.1 INTRODUCTION

Influenza viruses infect various species, including humans, and occasionally cause worldwide pandemics and severe outbreaks in poultry. The known types of influenza are A, B, C, D, and unclassified toad and fish influenza viruses (McCauley et al. 2012; Hause et al. 2014; Shi et al. 2018). Influenza A viruses are further classified into subtypes defined by the antigenic properties of the surface glycoproteins hemagglutinin (HA) and neuraminidase (NA), H1, H2, etc. and N1, N2, etc., respectively. The nomenclature for virus strains includes the type, the host (if non-human), geographical point of isolation, strain number, year, and subtype classification in parenthesis, e.g., A/Goose/Guangdong/1/1996(H5N1).

Influenza viruses have segmented single-stranded negative-sense RNA genomes consisting of seven (influenza C and D) or eight (influenza A and B) virus RNA (vRNA) segments. Each segment is bound to an heterortrimeric polymerase through their partially complementary 5′ and 3′ ends, the polymerase promoter, and the rest of the RNA is bound by multiple molecules of virus nucleoprotein (NP), forming the virus ribonucleoproteins (vRNPs). During the virus life cycle, RNAs of positive polarity are synthesized: complementary RNAs (cRNAs), which serve as templates for the synthesis of new vRNAs, and messenger RNAs (mRNAs), which are translated. Each of the segments encodes at least one essential protein. Higher-order structures formed by both negative- and positive-sense RNAs of influenza viruses have a number of important functions that regulate the replication of influenza viruses (reviewed by Gultyaev et al. 2010; Ferhadian et al. 2018).

The most conserved terminal "panhandle" promoter structures of influenza virus RNA segments have been extensively studied experimentally since the 1980s (Ferhadian et al. 2018). In contrast, functional RNA secondary structures folded in the internal parts of the segments were identified much later only by theoretical predictions (Gultyaev et al. 2007; Ilyinskii et al. 2009; Moss et al. 2011), followed by their confirmation by in vitro structure probing and mutagenesis experiments. The native secondary structures of influenza A virus vRNAs and mRNAs have been first studied in detail just recently (Dadonaite et al. 2019; Simon et al. 2019). This chapter discusses the bioinformatic approaches used in theoretical predictions and high-throughput studies of functional RNA structures in influenza virus genomes.

6.2 DETECTION OF CONSERVED STRUCTURES IN INFLUENZA VIRUS RNA BY COMPARATIVE ANALYSIS

The strategies for the prediction of functional virus RNA structures, in particular, in the absence of sufficient experimental data, usually exploit comparative analysis of secondary structures yielded by the algorithms based on the folding free energy minimization (reviewed by Schroeder 2009; Moss and Steitz 2015; Lim and Brown 2018). While the models based on straightforward calculations of the lowest free energy conformations considerably deviate from native structures of large RNA molecules, analysis of conserved domains folded in diverse sequences of related viruses or virus isolates can identify most likely functional motifs. The relatively fast evolution of influenza viruses continuously generates large numbers of new strains, and many thousands of influenza virus genomic sequences are available in GenBank and specialized databases (Bao et al. 2008; Shu and McCauley 2017). Predictions of influenza virus RNA folding can efficiently use this rich information in order to identify functional structured motifs.

The strategies for the search of conserved RNA structures in the influenza virus genome data usually include two main steps: the generation of models of thermodynamically stable structures and their validation by comparisons of suggested RNA conformations in different (groups of) strains. The first step may use folding simulations or free energy minimization of full-length RNA segments of representative virus strains (Gultyaev et al. 2007) or in sliding windows of 150–300 nucleotides along the sequences (Kobayashi et al. 2016). In order to retrieve conserved motifs more efficiently, the algorithms for the prediction of optimal consensus RNA secondary structures in sequence alignments, RNAalifold and RNAz, have been extensively applied for influenza virus RNAs in the last decade (Moss et al. 2011; Dela-Moss et al. 2014; Gultyaev et al. 2014, 2016; Soszynska-Jozwiak et al. 2015; Kobayashi et al. 2016, 2018). The RNAalifold algorithm (Bernhart et al. 2008) computes the low free energy structure possible in all aligned sequences. RNAz (Gruber et al. 2010) detects the most likely structured domains in overlapping windows of the alignment, scoring them on the basis of their conservation and free energy deviations from the values in random sequences of the same composition. The datasets of aligned influenza virus RNA sequences, used for consensus structure predictions, should be constructed with sequences that maximally cover their diversity in the explored phylogenetic range:

e.g., with representative strains of different subtypes, from different hosts, and geographic areas.

Prediction of RNA pseudoknots requires the application of special folding algorithms. Most of the free energy minimization algorithms do not consider pseudoknot formation due to computational difficulties, in particular in case of noncanonical pseudoknot topologies conserved in influenza virus mRNAs (Gultyaev et al. 2007; Moss et al. 2011). These pseudoknot structures have been predicted in NS, M, and PB1 mRNAs using heuristic algorithms that simulate stepwise pseudoknot folding (Gultyaev 1991; Xayaphoummine et al. 2005; Sperschneider and Datta 2010). A functional pseudoknot in the NP vRNA packaging region has been found using manual inspection of the base-pairing potential in the hairpin loops predicted by the RNAalifold algorithm, followed by estimates of pseudoknot free energies (Gultyaev et al. 2014).

Due to limitations of the used algorithms, their neglecting of NP binding effects, and the absence of conserved conformations over complete lengths of influenza virus RNAs, structure predictions obtained on either individual sequences or alignments only partially reproduce the native conformations. Nevertheless, such predictions can suggest candidate local conserved domains for subsequent analysis. Criteria for validation of such domains using sequence data may include their presence and thermodynamic stability in the predictions for distant representative strains (Gultyaev et al. 2007), relative significance of RNAz scores (Moss et al. 2011; Dela-Moss et al. 2014; Soszynska-Jozwiak et al. 2015; Kobayashi et al. 2016; Gultyaev et al. 2016), suppression of synonymous codon usage (Moss et al. 2011; Kobayashi et al. 2016, 2018), repeated presence in the structural models generated from datasets with randomly chosen strains covering the same phylogenetic range (Gultyaev et al. 2014, 2016; Soszynska-Jozwiak et al. 2015; Kobayashi et al. 2016, 2018), the presence of substitutions preserving base-pairing, and, in particular, covariations (Gultyaev et al. 2007, 2014, 2016; Moss et al. 2011). Such approaches identified a number of local structures conserved in all or some subtypes of influenza A (Figure 6.1) and a few structures in influenza B, C, and D viruses. Smaller numbers of conserved structured domains predicted in the latter types as compared to influenza A are mostly determined by a lower diversity of sequences that limits the reliability of comparative structural analysis.

Whatever strategy used for obtaining a structural RNA model, the presence of significant covariations in the thermodynamically reasonable structural elements is the most reliable evidence possible to obtain

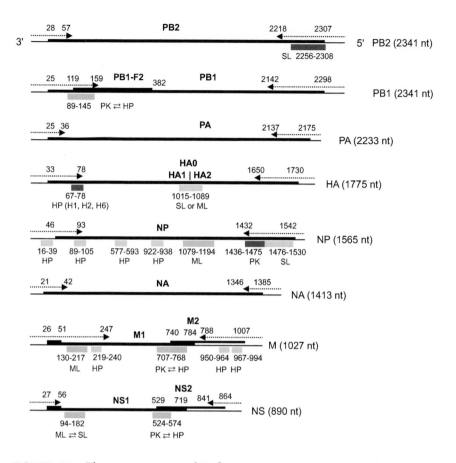

FIGURE 6.1 The most conserved influenza A virus RNA secondary structures, predicted by comparative analysis (Gultyaev et al. 2007, 2010, 2014, 2016; Ilyinskii et al. 2009; Moss et al. 2011; Soszynska-Jozwiak et al. 2015; Kobayashi et al. 2016, 2018). Nucleotide positions are indicated for A/Puerto Rico/8/1934 (H1N1). The extensions of minimal packaging signals from the segment termini (dashed arrows) are taken from Gerber et al. (2014). The open-reading frames of the main virus proteins are indicated above the lines showing the segments. The regions of predicted structured domains are indicated below by colored bars, red—structures predicted in vRNA, green—mRNA structures, magenta—structures predicted in RNAs of both polarities. HP, hairpin; SL, stem-loop structure; ML, multiloop branched structure; PK, pseudoknot.

from sequence data alone (e.g., Rivas et al. 2020). Despite the presence of sequences from thousands of diverse strains of influenza viruses in the databases, a mere conservation of a base-paired stem in multiple strains is not sufficient for the conclusion on its actual folding in RNAs of these viruses, because the stem nucleotides can be conserved for reasons other than RNA conformation requirements. Covariations have been found indeed in a number of predicted influenza A virus RNA structures, in particular, in NS, M, NP, and HA segments (Gultyaev et al. 2007, 2010, 2014, 2016; Moss et al. 2011; Spronken et al. 2017). A number of these covariations are characterized by high values of mutual information normalized by positional entropies, suggesting significant constraints imposed by RNA folding on the virus genome evolution (Gultyaev et al. 2014, 2016). It should be noted, however, that folding simulations using permuted influenza A virus RNA sequences have shown that high mutual information values of single covariations are not sufficient to prove structural models, while multiple covariation events in the same base pair and/or structure can be considered as significant evidence despite low mutual information (Gultyaev et al. 2016). On the other hand, the absence of significant covariation support for a specific structure does not mean that it does not exist. Some structures are formed by strongly conserved sequences, or, the opposite, are folded only in certain strains. Furthermore, a number of predicted structured domains in the influenza virus RNAs have been shown to be functional.

RNA secondary structures, including pseudoknots, near the splice sites of M and NS mRNAs (Figure 6.1), have been among the first conserved structures predicted in the coding regions of influenza virus RNAs, and their locations indicated an important role in the regulation of alternative splicing (reviewed by Dubois et al. 2014). In influenza A viruses, multibranch or stem-loop structures downstream of 5′ splice sites and equilibria between pseudoknot and hairpin conformations in the 3′ splice site regions have been predicted (Gultyaev et al. 2007, 2010; Ilyinskii et al. 2009; Moss et al. 2011, 2012). Folding of these structures in vitro has been supported by chemical and enzymatic probing and NMR spectroscopy (Gultyaev et al. 2007; Priore et al. 2013a; Jiang et al. 2014). Mutations in the splice site regions of M and NS mRNAs, designed with maximizing the effects on ensembles of possible structures, have been shown to change the ratios of spliced and unspliced RNAs and reduce virus titers in cell cultures (Jiang et al. 2016). Moreover, the compensatory mutant with restored pseudoknot/hairpin equilibrium at the 3′ splice site of NS segment had also the restored wild type splicing pattern and virus titer. This equilibrium can be important for adaptation of splicing regulation; for instance, it is shifted in favor of hairpin in the lineage of highly pathogenic avian influenza H5N1 viruses, which are circulating worldwide (Gultyaev et al. 2007). It has also been suggested that the changes in the structures predicted upstream of the NS 3′ splice site determine neurovirulent phenotype (Ward et al. 1995). Mutations in the conserved structure downstream of the 5′ splice site in the NS gene intron have been shown to affect the expression of NS1 protein early in infection (Baranovskaya et al. 2019).

Surprisingly, a hairpin predicted in the M segment packaging signal region has been found to influence splicing, even though it is located more than 200 nucleotides away from the 3′ end of the M gene intron (Spronken et al. 2017) (Figure 6.1). Mutations disrupting this hairpin led to changes in splicing, whereas hairpin-restoring compensatory mutants also restored the wild-type splicing pattern. Possibly, alternative structures in the disrupted hairpin mutants are formed that interfere with 3′ splice site recognition.

Despite considerable differences between the sequences of NS segments from influenza A and B viruses, the regions encompassing their 3′ splice sites can be folded into similar pseudoknots (Gultyaev et al. 2007; Gultyaev and Olsthoorn 2010). Structures predicted around other splice sites in the NS and M segments of influenza B and C viruses are conserved within a given type only (Dela-Moss et al. 2014) and have not been experimentally verified yet.

A few conserved structures have been predicted in the so-called minimal packaging signal regions of M, NP, PB2, and HA segments of influenza A viruses (Gultyaev et al. 2010, 2014, 2016; Kobayashi et al. 2016, 2018; Spronken et al. 2017) (Figure 6.1). Apart from universally conserved panhandle structures at the very ends of all segments, the minimal packaging signal regions include their untranslated parts, extending further into the coding parts by different lengths. Functional importance of the M, NP, and PB2 structures in these regions is supported by effects of structure-destabilizing mutations in mutant viruses (Gultyaev et al. 2014; Kobayashi et al. 2016, 2018; Spronken et al. 2017; Williams et al. 2018; Takizawa et al. 2019, 2020). The folding of a conserved hairpin in this region of M segment has been shown to play a role in virus replication by compensatory mutagenesis (Spronken et al. 2017). The compensatory mutagenesis of two adjacent conserved structures in the NP packaging signal region, a pseudoknot and a stem-loop, has suggested the pseudoknot to be important for virus replication, while no effects have been observed upon the stem-loop disruption (Gultyaev et al. 2014). An experimental study has demonstrated that the pseudoknot structure reduces NP protein binding in this region of NP segment, facilitating intersegmental interactions required for coordinated genome packaging into virions (Williams et al. 2018). The role of this pseudoknot in segment "bundling" has also been shown by mapping of RNA-RNA interactions (Takizawa et al. 2020).

In contrast to M, NP, and PB2 packaging signal region structures, which are conserved in all known subtypes of influenza A viruses, only subtype-specific structures have been predicted in the minimal packaging signal regions of HA segments apart from universally conserved panhandle (Gultyaev et al. 2016). Relatively low mutual information values in the detected covariations indicate weak structural constraints, and disruption of these structures in mutant viruses has not revealed any effect on virus fitness (Gultyaev et al. 2016). This might be explained by redundancy of some packaging signal motifs (Fujii et al. 2009; Bolte et al. 2019). Interestingly, the negative effect on replication in human cell culture has been observed for mutant viruses with stabilized hairpin structure that is conserved only in HA packaging signals of swine strains (Canale et al. 2018).

Covariations in a number of local structures predicted in the central regions of NP and HA positive- and negative-sense RNAs of influenza A viruses indicate the presence of RNA structural constraints (Gultyaev et al. 2014, 2016). Similar to the minimal packaging signals of HA segment, its central domains can form subtype-specific structures which are conserved only in one or a few related subtypes. This is consistent with the subtype-specific pattern of codon conservation, which defines motifs important for segment packaging (Gog et al. 2007). Structures in the segment interiors could be important for interactions between segments upon vRNA packaging, presumably mediated by kissing intersegmental loop-loop contacts (Hutchinson et al. 2010; Noda et al. 2012; Gavazzi et al. 2013a; Shafiuddin and Boon, 2019). On the other hand, the observed structural constraints may be determined by the importance of local mRNA structures, as disruptions of some of them attenuate virus replication due to yet undetermined functions such as modulation of cotranslational protein folding or maintenance of transcript stabilities (Gultyaev et al. 2016; Simon et al. 2019).

The most conserved structured domain in HA segments, supported by a number of covariation events, has been identified in the region encompassing the boundary between the parts encoding two polypeptide chains, HA1 and HA2, which are produced by cleavage of the polyprotein HA0 (Gultyaev et al. 2016). In particular, H5- and H7-specific stem-loop structures have been suggested to play a role in evolving toward highly pathogenic avian influenza strains due to insertions of codons into the hairpin loops that result in multibasic cleavage site motifs. On the other hand, similar RNA structures are conserved in some other subtypes of influenza A and even in influenza B and C viruses, which are not known to evolve toward highly pathogenic strains. This suggests that the structured character of this domain could also be important for other, yet unknown, function(s) (Gultyaev et al. 2019). Interestingly, this is the only structured domain in the HA segment of influenza A viruses, predicted by the segment-wide scanning based on folding free energy and codon conservation statistics (Moss et al. 2011).

Covariation analysis and mutagenesis experiments with some of the predicted structures in NP and HA RNAs suggest that a number of local structured domains in vRNAs, mRNAs, and/or cRNAs may contribute to the virus fitness without being individually essential for the virus replication (Gultyaev et al. 2014, 2016). The global ordered RNA structure (GORS), predominantly manifested in deviation of potential folding free energy from the values in comparable random sequences, has been detected in influenza A and B viruses indeed (Priore et al. 2012, 2013b). Furthermore, the differences between RNA folding stabilities of the influenza A viruses isolated from different hosts have been noted (Brower-Sinning et al. 2009; Priore et al. 2012). Possible explanation for these differences is the adaptation to the host's body temperature and pH in the infected cells. Temperature-sensitive local structures in virus mRNA have been identified by folding free energy minimization at different temperatures, showing correlations with substitutions in cold-sensitive strains (Chursov et al. 2012).

6.3 IDENTIFICATION OF INFLUENZA VIRUS RNA STRUCTURES USING STRUCTURE PROBING

Predictions of fragmentary local structured domains by comparative RNA analysis (Figure 6.1) leave a large portion of the influenza virus genome where no reliable prediction can be made either because the sequences are too conserved for the detection of structural constraints or the structures differ between diverse virus strains. Structure predictions in such regions can be done using RNA experimental structure probing data which provide the information on probabilities of particular nucleotides to be paired. The basic idea of such prediction is folding free energy minimization constrained by the probing data (Mathews et al. 2004; Deigan et al. 2009; Washietl et al. 2012). Options for the incorporation of probing data are implemented in the widely used programs for RNA structure prediction (Lorenz et al. 2011; Bellaousov et al. 2013).

Using the data from various probing methods, the in vitro structures of NP, M, and NS vRNA and of complementary NP RNA (cRNA) from a highly pathogenic H5N1 strain A/Vietnam/1203/2004 (H5N1) have been predicted (Soszynska-Jozwiak et al. 2017; Ruszkowska et al. 2016; Lenartowicz et al. 2016; Michalak et al. 2019). With a few exceptions, conserved structures suggested by comparative analysis (Figure 6.1) have been mostly recapitulated in these models. In addition, a number of other structured domains have been identified, some of them being preserved in multiple strains due to sequence conservation and/or substitutions that do not disrupt the predicted RNA structures.

Structure probing-guided predictions of the A/Puerto Rico/8/1934 (H1N1) mRNA structures in vivo show that most of the virus mRNA folding potential is not realized in the cytoplasm of infected cell: while almost half of the virus transcriptome is folded in vitro, more than 80% is unfolded in vivo (Simon et al. 2019). Such a difference is typical for cellular mRNA, being mostly explained by mRNA unfolding by translating ribosomes. Nevertheless, some locally stable short-range interactions are persistent in mRNA in vivo. Moreover, disruption of several such structures resulted in slower replication kinetics of mutant viruses (Simon et al. 2019). However, conserved pseudoknots encompassing the splice sites of M and NS mRNAs (Figure 6.1), shown to regulate their splicing (Jiang et al. 2016), have not been found in cytoplasm environment, suggesting that these mRNAs fold differently in nucleus (Simon et al. 2019).

Native structures of influenza A virus genomic vRNAs in vRNP complexes have recently been probed using selective 2′-hydroxyl acylation analyzed by primer extension and mutational profiling (SHAPE-MaP) data (Dadonaite et al. 2019). These models provide rich information on structure-function relationships in vRNAs, which exert their functions in vRNP complexes rather than in the "naked" states. As expected from the melting effects of NP binding on vRNA (Baudin et al. 1994), SHAPE-MaP probing shows that native vRNAs are less structured than allowed by their base-pairing potential, but they still form a substantial number of local structures (Dadonaite et al. 2019). Conserved structured domains, predicted in vRNA segments by comparative analysis (Figure 6.1), are mostly consistent with the vRNP models derived from the SHAPE-MaP data. On the other hand, the differences in vRNP conformation are predicted for different strains of influenza A viruses, in particular, for those of different subtypes (Dadonaite et al. 2019). Interestingly, these differences are not restricted to HA and NA segments, which define subtype nomenclature.

Locations of stable RNA secondary structures in vRNP particles correlate with the regions of relatively low NP binding, while SHAPE probing shows NP-enriched regions to be less structured (Williams et al. 2018; Dadonaite et al. 2019). Recent studies have revised the "beads on a string" model of NP binding to vRNA in influenza A and B viruses, showing nonuniform and nonrandom association of protein molecules (Lee et al. 2017; Williams et al. 2018; Le Sage et al. 2018). Viruses of the same subtype have similar, but not identical, nonperiodic NP binding profiles with alternating regions of abundant and poor binding, and some similarities have been observed between H1N1 and H3N2 influenza A viruses, but no similarity could be found between influenza A and B viruses (Le Sage et al. 2018). Low-NP binding regions seem to provide an opportunity for RNA to form intra- and/or intermolecular base pairs, and locally stable intramolecular structures can be predicted by free energy minimization in about half of the NP-free regions in the A/Puerto Rico/8/1934 (H1N1) (Williams et al., 2018).

6.4 NETWORKS OF INTERSEGMENTAL INTERACTIONS

Local RNA structures and/or NP-free regions in influenza virus vRNP segments are thought to be important for intersegmental interactions that determine coordinated packaging of vRNPs into virions (Hutchinson et al. 2010; Noda et al. 2012; Gerber et al. 2014; Shafiuddin and Boon 2019). Electron tomography of vRNP complexes has revealed multiple interactions between RNA segments, and measurements of interactions between naked influenza A virus genomic segments *in vitro* have suggested the existence of subtype-specific interaction networks (Fournier et al. 2012; Noda et al. 2012; Gavazzi et al. 2013b). Recent developments in the methods for the identification of RNA-RNA contacts using proximity ligation (Aw et al. 2016; Sharma et al. 2016; Lu et al. 2016) opened the ways for mapping such networks *in virio*. The mapping is based on the crosslinking of closely located RNA fragments using psoralen or a psoralen derivative, followed by partial RNA digestion, ligation of crosslinked fragments, and massive parallel sequencing. In order to better preserve the network of contacts formed by vRNAs with bound nucleoprotein molecules, extra steps of UV-induced cross-linking and immunoprecipitation have been suggested (Le Sage et al. 2020). Analysis of chimera reads yields interacting regions.

Crosslinking-based methods detect hundreds of contacts between influenza A virus RNA segments in virions (Dadonaite et al. 2019; Le Sage et al. 2020). These contacts suggest that vRNP packaging occurs via complex flexible networks of intersegmental interactions, which involve all parts of the segments (Figure 6.2a). Predicted base-pairing structures formed by interacting regions have lower median free energies than those predicted with reshuffled sequences of the partners, confirming that the interactions identified by cross-linking are indeed determined by specific intermolecular stem-loops (Dadonaite et al. 2019). Some contacts are formed more frequently, and some vRNA regions can interact with multiple sites in other segments. On the other hand, substitutions at such hotspots (Figure 6.2b) do not necessarily lead to fitness loss, as alternative hotspots could create a functional rearranged network (Le Sage et al. 2020). Due to redundancy of many contacts, different conformations are possible even for genetically identical combinations of vRNA segments, and the differences between various influenza A viruses grow with sequence dissimilarity.

Similar to intramolecular folding predictions, predictions of the most likely intermolecular base-paired structures can be done using free energy minimization (e.g. Krüger and Rehmsmeier 2006; Mann et al. 2017). Applied for intersegmental contacts identified by crosslinking, such predictions yield the boundaries of the interacting regions that form energetically favorable intermolecular structures (Dadonaite et al. 2019; Le Sage et al. 2020) (Figure 6.2b). The formation of extended intermolecular duplexes can be facilitated by transient kissing interactions between the hairpin loops formed by the interacting regions, followed by hairpin disruption upon duplex propagation with increasing numbers of intermolecular base pairs (Brunel et al. 2002). Such kissing hairpins have been predicted to initiate an interaction between segments 2 and 8 of avian H5N2 strains, first detected in the *in vitro* vRNA interaction network and then shown by compensatory mutagenesis to promote copackaging of vRNA segments in viral particles (Gavazzi et al. 2013a,b) (Figure 6.2c).

The mechanism of coordinated copackaging of vRNP segments, determined by specific interaction networks, has important implications for gene reassortment (Essere et al. 2013; Gavazzi et al. 2013a; Gerber et al.

2014; Dadonaite et al. 2019). Certain interactions are critical for packaging, and the lack of their conservation might limit reassortment between distant strains of influenza A viruses (Essere et al. 2013; Gavazzi et al. 2013a). Studies on the interaction networks can reveal specific intersegmental contacts that are required for a particular segment constellation to be compatible. For instance, the interaction between the regions in PB1 and N2 NA segments originating from H3N2 strains (Figure 6.2d) has been shown to be important in the reassortment with H1N1 viruses (Dadonaite et al. 2019). Interestingly, this observation correlates with co-occurrence of avian PB1 and N2 NA segments in two pandemics of the past century and cosegregation of these segments, when derived from H3N2 strains, in studies on reassortant seed vaccine viruses.

The assembly of eight influenza A vRNPs into a supramolecular structure is a dynamic process that occurs via intermediate interactions of increasing numbers of vRNPs en route to the cell membrane. Possible sequences of these intermediate events have recently been modeled using the data on binary segment colocalization and point process models (Majarian et al. 2019). In these models, the data on segment colocalizations in pairs, triplets, and quadruplets have been used to construct the graphs with the nodes corresponding to sets of vRNP segments. The optimal path from single nodes to the complete node with all 8 vRNPs could be reconstructed by a dynamic programming algorithm. This yielded the most likely assembly scenario: first formation of two separate complexes with HA, M, NA, NS and PB1, PB2, NP, respectively, that finally merge with the PA segment. It should be noted, however, that this model does not exclude multiple interaction pathways that may be realized in parallel, especially because the modeling is based on colocalization data obtained at one time point. Potential incorporation of spatiotemporal modeling in this approach could make it more informative (Majarian et al. 2019).

6.5 CONCLUDING REMARKS

High-throughput experimental approaches, which allow the accumulation of rich information on vRNP structures, formulate a number of new challenging problems in bioinformatic analysis of the complex interplay of higher-order structure formation, RNA-RNA interactions, and NP binding in the assembly and functioning of influenza vRNP particles. For instance, intersegmental interactions are realized via base-pairing of rather extended regions in the segments (Figure 6.2), which is expected to compete with individual segment folding. On the other hand, such interactions

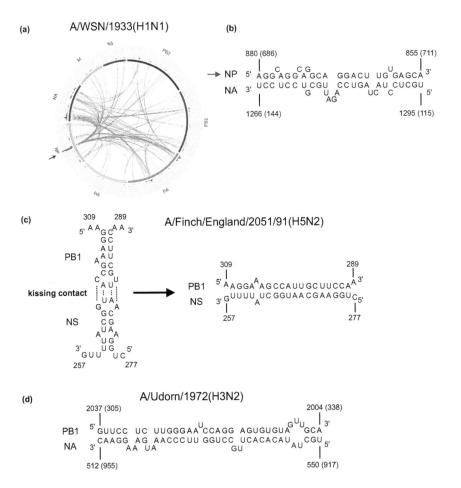

FIGURE 6.2   Examples of intersegmental RNA-RNA interactions in the influenza A virus strains. (a) The A/WSN/1933(H1N1) interaction network. (Reproduced from Le Sage et al. (2020), available under Creative Commons Attribution 4.0 International License, http://creativecommons.org/licenses/by-nc-nd/4.0/) The top 25% of interactions is shown. The outer circle histogram shows the relative frequency of intersegmental interactions at a given site. A major interaction hotspot in the NP segment is indicated by red arrow. (b) Predicted RNA duplex formed between the NP hotspot and one of NA regions (Le Sage et al. 2020). (c) The interaction between PB1 and NS segments, required for optimal replication of A/Finch/England/2051/91(H5N2), and predicted kissing contacts between initiating hairpins (Gavazzi et al. 2013a). (d) A prominent interaction between PB1 and NA segments in the interaction network of A/Udorn/1972(H3N2) (Dadonaite et al. 2019). The standard nucleotide numbering of influenza virus negative-sense vRNA segments in the 3′-5′ direction is used; position numbers in the 5′-3′ direction from crosslinking experiments in (b) and (d) are indicated in brackets.

are likely to be initiated by contacts between the loops formed by the local structures in the segments, which should be disrupted upon propagation of intersegmental pairing along the sequences (Gavazzi et al. 2013a). Apparently, structure predictions, focused on the regions identified to be involved in the RNA-RNA interactions, should take into account dynamic equilibria between alternative conformations. This requires the processing of the structure predictions that yield probabilities of all potential base pairs across the virus genome, with comparisons of results obtained for different virus strains both in virio and with naked vRNAs. The first results in this direction, which combine computational approaches with structure probing data (Dadonaite et al. 2019), are very promising.

The long-standing problem of properly taking into account the effects of protein binding in the predictions of native RNA structures is very much pronounced in the case of influenza virus RNAs. Due to the melting effects of multiple NP molecules, straightforward free energy minimization makes sense only for short-range local structures in vRNAs, which are likely to be NP-free or reliably supported by comparative analysis. On the other hand, NP binding itself may be determined by stable local vRNA folding (Williams et al. 2018).

Due to the differences between the subclades and even individual strains, comparative analysis, based on structural constraints detectable in sequence data, can identify only a minor part of influenza virus RNA structurome. The remaining domains form clade- or even strain-specific structures that can be determined by combining bioinformatic approaches with high-throughput experimental methods. Both conserved and variable parts determine a number of important functions in virus replication and are driving forces in virus evolution. The knowledge of conserved structured domains helps to understand the structural constraints in the influenza genome evolution, while the information on unique functional vRNA structures allows one to predict the constraints in gene reassortment.

REFERENCES

Aw, J.G.A., Shen, Y., Wilm, A. et al. 2016. *In vivo* mapping of eukaryotic RNA interactomes reveals principles of higher-order organization and regulation. *Mol. Cell* 62:603–17.

Bao, Y., Bolotov, P., Dernovoy, D., et al. 2008. The influenza virus resource at the National Center for Biotechnology Information. *J. Virol.* 82:596–601.

Baranovskaya, I., Sergeeva, M., Fadeev, A., et al. 2019. Changes in RNA secondary structure affect NS1 protein expression during early stage influenza virus infection. *Virol. J.* 16:162.

Baudin, F., Bach, C., Cusack, S., and Ruigrok, R.W.H. 1994. Structure of influenza virus RNP. I. Influenza virus nucleoprotein melts secondary structure in panhandle RNA and exposes the bases to the solvent. *EMBO J.* 13:3158–65.

Bellaousov, S., Reuter, J.S., Seetin, M.G., and Mathews, D.H. 2013. RNAstructure: Web servers for RNA secondary structure prediction and analysis. *Nucleic Acids Res.* 41:W471–4.

Bernhart, S.H., Hofacker, I.L., Will, S., Gruber, A.R., and Stadler, P.F. 2008. RNAalifold: Improved consensus structure prediction for RNA alignments. *BMC Bioinf.* 9:474.

Bolte, H., Rosu, M.E., Hagelauer, E., García-Sastre, A., and Schwemmle, M. 2019. Packaging of the influenza virus genome is governed by a plastic network of RNA- and nucleoprotein-mediated interactions. *J. Virol.* 93:e01861–18.

Brower-Sinning, R., Carter, D.M., Crevar, C.J., Ghedin. E., Ross, T.M., and Benos, P.V. 2009. The role of RNA folding free energy in the evolution of the polymerase genes of the influenza A virus. *Genome Biol.* 10:R18.

Brunel, C., Marquet, R., Romby, P., and Ehresmann, C. 2002. RNA loop-loop interactions as dynamic functional motifs. *Biochimie* 84:925–44.

Canale, A.S., Venev, S.V., Whitfield, T.W. et al. 2018. Synonymous mutations at the beginning of the influenza A virus hemagglutinin gene impact experimental fitness. *J. Mol. Biol.* 430:1098–115.

Chursov, A., Kopetzky, S.J., Leshchiner, I., et al. 2012. Specific temperature-induced perturbations of secondary mRNA structures are associated with the cold-adapted temperature-sensitive phenotype of influenza A virus. *RNA Biol.* 9:1266–74.

Dadonaite, B., Gilbertson, B., Knight, M.L., et al. 2019. The structure of the influenza A virus genome. *Nat. Microbiol.* 4:1781–9.

Deigan, K.E., Li, T.W., Mathews, D.H., and Weeks, K.M. 2009. Accurate SHAPE-directed RNA structure determination. *Proc. Natl. Acad. Sci. U. S. A.* 106:97–102.

Dela-Moss, L.I., Moss, W.N., and Turner, D.H. 2014. Identification of conserved RNA secondary structures at influenza B and C splice sites reveals similarities and differences between influenza A, B, and C. *BMC Res. Notes* 7:22.

Dubois, J., Terrier, O., and Rosa-Calatrava, M. 2014. Influenza viruses and mRNA splicing: Doing more with less. *mBio* 5:e00070–14.

Essere, B., Yver, M., Gavazzi, C., et al. 2013. Critical role of segment-specific packaging signals in genetic reassortment of influenza A viruses. *Proc Natl Acad. Sci. U. S. A.* 110:E3840–8.

Ferhadian, D., Contrant, M., Printz-Schweigert, A., Smyth, R.P., Paillart, J.C., and Marquet, R. 2018. Structural and functional motifs in influenza virus RNAs. *Front. Microbiol.* 9:559.

Fournier, E., Moules, V., Essere, B., et al. 2012. A Supramolecular assembly formed by influenza A virus genomic RNA segments. *Nucleic Acids Res.* 40:2197–209.

Fujii, K., Ozawa M., Iwatsuki-Horimoto, K., Horimoto, T., and Kawaoka, Y. 2009. Incorporation of influenza A virus genome segments does not absolutely require wild-type sequences. *J. Gen. Virol.* 90:1734–40.

Gavazzi, C., Isel, C., Fournier, E., et al. 2013b. An *in vitro* network of intermolecular interactions between viral RNA segments of an avian H5N2 influenza A virus: Comparison with a human H3N2 virus. *Nucleic Acids Res.* 41:1241–54.

Gavazzi, C., Yver, M., Isel, C., et al. 2013a. A functional sequence-specific interaction between influenza A virus genomic RNA segments. *Proc. Natl. Acad. Sci. U. S. A.* 110:16604–9.

Gerber, M., Isel, C., Moules, V., and Marquet, R. 2014. Selective packaging of the influenza A genome and consequences for genetic reassortment. *Trends Microbiol.* 22:446–55.

Gog, J.R., Dos Santos Afonso, E., Dalton, R.M., et al. 2007. Codon conservation in the influenza A virus genome defines RNA packaging signals. *Nucleic Acids Res.* 35:1897–907.

Gruber, A.R., Findeiß, S., Washietl, S., Hofacker, I.L., and Stadler, P.F. 2010. RNAz 2.0: Improved noncoding RNA detection. *Pac. Symp. Biocomput.* 2010:69–79.

Gultyaev, A.P. 1991. The computer simulation of RNA folding involving pseudoknot formation. *Nucleic Acids Res.* 19:2489–94.

Gultyaev, A.P., and Olsthoorn, R.C.L. 2010. A family of non-classical pseudoknots in influenza A and B viruses. *RNA Biol.* 7:125–9.

Gultyaev, A.P., Heus, H.A., and Olsthoorn, R.C.L. 2007. An RNA conformational shift in recent H5N1 influenza A viruses. *Bioinformatics* 23:272–6.

Gultyaev, A.P., Fouchier, R.A.M., and Olsthoorn, R.C.L. 2010. Influenza virus RNA structure: Unique and common features. *Int. Rev. Immunol.* 29:533–56.

Gultyaev, A.P., Tsyganov-Bodounov, A., Spronken, M.I., van der Kooij, S., Fouchier, R.A.M., and Olsthoorn, R.C.L. 2014. RNA structural constraints in the evolution of the influenza A virus genome NP segment. *RNA Biol.* 11:942–52.

Gultyaev, A.P., Spronken, M.I., Richard, M., Schrauwen, E.J., Olsthoorn, R.C.L., and Fouchier, R.A.M. 2016. Subtype-specific structural constraints in the evolution of influenza A virus hemagglutinin genes. *Sci. Rep.* 6:38892.

Gultyaev, A.P., Richard, M., Spronken, M.I., Olsthoorn, R.C.L., and Fouchier, R.A.M. 2019. Conserved structural RNA domains in regions coding for cleavage site motifs in hemagglutinin genes of influenza viruses. *Virus Evol.* 5:vez034.

Hause, B.M., Collin, E.A., Liu, R., et al. 2014. Characterization of a novel influenza virus in cattle and swine: Proposal for a new genus in the *Orthomyxoviridae* family. *mBio* 5:e00031–14.

Hutchinson, E.C., von Kirchbach, J.C., Gog, J.R., and Digard, P. 2010. Genome packaging in influenza A virus. *J. Gen. Virol.* 91:313–28.

Ilyinskii, P.O., Schmidt, T., Lukashev, D., et al. 2009. Importance of mRNA secondary structural elements for the expression of influenza virus genes. *OMICS* 13:421–30.

Jiang, T., Kennedy, S.D., Moss, W.N., Kierzek, E., and Turner, D.H. 2014. Secondary structure of a conserved domain in an intron of influenza A M1 mRNA. *Biochemistry* 53:5236–48.

Jiang, T., Nogales, A., Baker, S.F., Martinez-Sobrido, L., and Turner, D.H. 2016. Mutations designed by ensemble defect to misfold conserved RNA structures of influenza A segments 7 and 8 affect splicing and attenuate viral replication in cell culture *PLoS One* 11:e0156906.

Kobayashi, Y., Dadonaite, B., van Doremalen, N., Suzuki, Y., Barclay, W.S., and Pybus, O.G. 2016. Computational and molecular analysis of conserved influenza A virus RNA secondary structures involved in infectious virion production. *RNA Biol.* 13:883–94.

Kobayashi, Y., Pybus, O.G., Itou, T., and Suzuki, Y. 2018. Conserved secondary structures predicted within the 5′ packaging signal region of influenza A virus PB2 segment. *Meta Gene* 15:75–9.

Krüger, J., and Rehmsmeier, M. 2006. RNAhybrid: MicroRNA target prediction easy, fast and flexible. *Nucleic Acids Res.* 34:W451–4.

Le Sage, V., Nanni, A.V., Bhagwat A.R., et al. 2018. Non-uniform and non-random binding of nucleoprotein to influenza A and B viral RNA. *Viruses* 10:522.

Le Sage, V., Kanarek, J.P., Snyder, D.J., Cooper, V.S., Lakdawala, S.S., and Lee, N. 2020. Mapping of influenza virus RNA-RNA interactions reveals a flexible network. *Cell Rep.* 31:107823.

Lee, N., Le Sage, V., Nanni, A.V., Snyder, D.J., Cooper, V.S., and Lakdawala, S.S. 2017. Genome-wide analysis of influenza viral RNA and nucleoprotein association. *Nucleic Acids Res.* 45:8968–77.

Lenartowicz, E., Kesy, J., Ruszkowska, A. et al. 2016. Self-folding of naked segment 8 genomic RNA of influenza A virus. *PLoS One* 11:e0148281.

Lim, C.S., and Brown, C.M. 2018. Know your enemy: Successful bioinformatic approaches to predict functional RNA structures in viral RNAs. *Front. Microbiol.* 8:2582.

Lorenz, R., Bernhart, S.H., Höner zu Siederdissen, C., et al. 2011. ViennaRNA package 2.0. *Algorithms Mol. Biol.* 6:26.

Lu, Z., Zhang, Q, C., Lee, B., et al. 2016. RNA duplex map in living cells reveals higher-order transcriptome structure. *Cell* 165:1267–79.

Majarian, T.D., Murphy, R.F., and Lakdawala, S.S. 2019. Learning the sequence of influenza A genome assembly during viral replication using point process models and fluorescence *in situ* hybridization. *PLoS Comput. Biol.* 15:e1006199.

Mann, M., Wright, P., and Backofen, R. 2017. IntaRNA 2.0: Enhanced and customizable prediction of RNA-RNA interactions. *Nucleic Acids Res.* 45:W435–9.

Mathews, D.H., Disney, M.D., Childs, J.L., Schroeder, S.J., Zuker, M., and Turner, D.H. 2004. Incorporating chemical modification constraints into a dynamic programming algorithm for prediction of RNA secondary structure. *Proc. Natl. Acad. Sci. U. S. A.* 19:7287–92.

McCauley, J.W., Hongo, S., Kaverin, N.V., et al. 2012. Orthomyxoviridae. In *Virus Taxonomy: Classification and Nomenclature of Viruses: Ninth Report of the International Committee on Taxonomy of Viruses*, ed. A.M.Q. King, M.J. Adams, E.B. Carstens, and E.J. Lefkowitz, pp. 749–61. Amsterdam: Elsevier Academic Press.

Michalak, P., Soszynska-Jozwiak, M., Biala, E., et al. 2019. Secondary structure of the segment 5 genomic RNA of influenza A virus and its application for designing antisense oligonucleotides. *Sci Rep.* 9:3801.

Moss, W.N., and Steitz, J.A. 2015. In silico discovery and modeling of non-coding RNA structure in viruses. *Methods* 91:48–56.

Moss, W.N., Priore, S.F., and Turner, D.H. 2011. Identification of potential conserved RNA secondary structure throughout influenza A coding regions. *RNA* 17:991–1011.

Moss, W.N., Dela-Moss, L.I., Kierzek, E., Kierzek, R., Priore, S.F., and Turner, D.H. 2012. The 3′ splice site of influenza A segment 7 mRNA can exist in two conformations: A pseudoknot and a hairpin. *PLoS One* 7:e328323.

Noda, T., Sugita, Y., Aoyama, K., et al. 2012. Three-dimensional analysis of ribonucleoprotein complexes in influenza A virus. *Nat. Commun.* 3:639.

Priore, S.F., Moss, W.N., and Turner, D.H. 2012. Influenza A virus coding regions exhibit host-specific global ordered RNA structure. *PLoS One* 7:e35989.

Priore, S.F., Kierzek, E., Kierzek, R., et al. 2013a. Secondary structure of a conserved domain in the intron of influenza A NS1 mRNA. *PLoS One* 8:e70615.

Priore, S.F., Moss, W.N., and Turner, D.H. 2013b. Influenza B virus has global ordered RNA structure in (+) and (−) strands but relatively less stable predicted RNA folding free energy than allowed by the encoded protein sequence. *BMC Res. Notes* 6:330.

Rivas, E., Clements, J., and Eddy, S.R. 2020. Estimating the power of sequence covariation for detecting conserved RNA structure. *Bioinformatics* 36:3072–6.

Ruszkowska, A., Lenartowicz, E., Moss, W.N., Kierzek, R., and Kierzek, E. 2016. Secondary structure model of the naked segment 7 influenza A virus genomic RNA. *Biochem J.* 473:4327–48.

Schroeder, S.J. 2009. Advances in RNA structure prediction from sequence: New tools for generating hypotheses about viral RNA structure-function relationships. *J. Virol.* 83:6326–34.

Shafiuddin, M., and Boon, A.C.M. 2019. RNA sequence features are at the core of influenza A virus genome packaging. *J. Mol. Biol.* 431:4217–28.

Sharma, E., Sterne-Weiler, T., O'Hanlon, D., and Blencowe, B.J. 2016. Global mapping of human RNA-RNA interactions. *Mol. Cell* 62:618–26.

Shi, M., Lin, X.D., Chen, X., et al. 2018. The evolutionary history of vertebrate viruses. *Nature* 556:197–202.

Shu, Y., and McCauley, J. 2017. GISAID: Global initiative on sharing all influenza data - from vision to reality. *Euro Surveill.* 22:30494.

Simon, L.M., Morandi, E., Luganini, A., et al. 2019. *In vivo* analysis of influenza A mRNA secondary structures identifies critical regulatory motifs. *Nucleic Acids Res.* 47:7003–17.

Soszynska-Jozwiak, M., Michalak, P., Moss, W.N., Kierzek, R., and Kierzek, E. 2015. A conserved secondary structural element in the coding region of the influenza A virus nucleoprotein (NP) mRNA is important for the regulation of viral proliferation. *PLoS One* 10:e0141132.

Soszynska-Jozwiak, M., Michalak, P., Moss, W.N., Kierzek, R., Kesy, J., and Kierzek, E. 2017. Influenza virus segment 5 (+)RNA: Secondary structure and new targets for antiviral strategies. *Sci. Rep.* 7:15041.

Sperschneider, J., and Datta, A. 2010. DotKnot: Pseudoknot prediction using the probability dot plot under a refined energy model. *Nucleic Acids Res.* 38:e103.

Spronken, M.I., van de Sandt, C.E., de Jongh, E.P., et al. 2017. A compensatory mutagenesis study of a conserved hairpin in the M gene segment of influenza A virus shows its role in virus replication. *RNA Biol.* 14:1606–16.

Takizawa, N., Ogura, Y., Fujita, Y., et al. 2019. Local structural changes of the influenza A virus ribonucleoprotein complex by single mutations in the specific residues involved in efficient genome packaging. *Virology* 531:126–40.

Takizawa, N., Higashi, K., Kawaguchi, R.K., et al. 2020. A functional structure in the influenza A virus ribonucleoprotein complex for segment bundling. *BioRxiv.* doi: 10.1101/2020.03.05.975870.

Ward, A.C., Azad, A.A., and McKimm-Breschkin, J.L. 1995. Changes in the NS gene of neurovirulent strains of influenza affect splicing. *Virus Genes* 10:91–4.

Washietl, S., Hofacker, I.L., Stadler, P.F., and Kellis, M. 2012. RNA folding with soft constraints: Reconciliation of probing data and thermodynamic secondary structure predictions. *Nucleic Acids Res.* 40:4261–72.

Williams, G.D., Townsend, D., Wylie, K.M., et al. 2018. Nucleotide resolution mapping of influenza A virus nucleoprotein-RNA interactions reveals RNA features required for replication. *Nat. Commun.* 9:465.

Xayaphoummine, A., Bucher, T., and Isambert, H. 2005. Kinefold web server for RNA/DNA folding path and structure prediction including pseudoknots and knots. *Nucleic Acids Res.* 33:W605–10.

CHAPTER 7

Structural Genomics and Interactomics of SARS-COV2

Decoding Basic Building Blocks of the Coronavirus

Ziyang Gao, Senbao Lu, Oleksandr Narykov, Suhas Srinivasan, and Dmitry Korkin

Worcester Polytechnic Institute

CONTENTS

7.1 UNDERSTANDING THE MOLECULAR MECHANISMS OF COVID-19: A CURRENT FOCUS OF SCIENTIFIC COMMUNITY

By November 2020, over 60 million people in 218 countries have been infected with COVID-19, with the number of deaths rapidly approaching 1.5 million and the second wave of pandemic underway in many countries [1]. The source of this devastating pandemic is SARS-CoV-2, a novel betacoronavirus originating from a bat coronavirus, with the original SARS-CoV being one of the closest relatives; other related viruses include MERS-CoV, a highly pathogenic virus, and HCoV-HKU1, which causes a common cold [2–5]. The key behind the pathogenicity and infectivity of SARS-CoV-2, SARS-CoV, and MERS-CoV lies in the functioning of their proteins. Much of this functioning is carried out at the molecular level through the interactions of these proteins with each other and with the host proteins [6]. On the other hand, to develop an effective treatment of the infection, one needs to understand the mechanistic nature of the interactions between the antiviral drugs and their targets—either viral proteins or host protein involved in the host-virus interactions [7,8]. Here, we discuss the role of computational approaches in uncovering structure and function of the basic protein building blocks of SARS-CoV-2, proposing next directions to understanding the virus and its functioning at the atomistic level.

7.2 GENOMIC AND STRUCTURAL ORGANIZATION OF THE NOVEL CORONAVIRUS

The SARS-CoV-2 genome and its organization are similar to other closely related betacoronaviruses, including human SARS-CoV, bat coronavirus (BtCoV), and murine coronavirus (Mouse hepatitis virus, MHV) [9,10]. The ~30 kb genome is predicted to encode 29 proteins encoded by as many as 14 open-reading frames (ORFs): (i) polyproteins encoded by ORF1a and ORF1ab are autoproteolytically processed into 16 predicted nonstructural proteins, Nsp1–Nsp16; (ii) at least 13 downstream open reading frames (ORFs) including four structural proteins, surface glycoprotein (Spike or S), envelope (E), membrane (M), and nucleocapsid (N); and (iii) nine accessory factors ORF3a, ORF3b, ORF6, ORF7a, ORF7b, ORF8, ORF9a, ORF9b, and ORF10. Multiple copies of the S-trimer, E-pentamer, and M-dimer protein complexes assemble together with the lipid bilayer molecules to form the viral envelope [11]. In contrast, the oligomeric complexes of another structural protein (N) interact with RNA, forming large ribonucleoprotein (RNP) structures, which still remain poorly understood. It has been recently estimated that there are roughly 30–35 RNPs per a single virion [11]. Functionally, many viral proteins have been characterized, including key enzymes: main chymotrypsin-like protease (M^{pro} or 3CL^{pro} or Nsp5), papain-like protease (PL^{pro} or Nsp3), helicase (Nsp13), and RNA-dependent RNA polymerase (RdRp or Nsp12).

7.3 STRUCTURAL CHARACTERIZATION OF THE INDIVIDUAL VIRAL PROTEINS

Understanding the structure of the proteins constituting the viral proteome is a critical first step in determining the function and evolution of the virus' basic components and holds the key to structural characterization of the virion particles (Figure 7.1). In spite of the joint efforts by the scientific community, we are yet to get a comprehensive structural characterization of all SARS-CoV-2 proteins using experimental methods. Currently, structures of 16 proteins, many of them only partially resolved, have been reported to PDB (Figure 7.1) according to Coronavirus3D database [12]. Furthermore, among the four structural proteins, only two, S and N, have been structurally resolved and only partially. Other proteins, such as Nsp3, consist of multiple domains, and while the experimental structure of each domain is available, it is not known how these domains form the overall 3D structure of the whole-length protein.

When an experimentally solved structure of a viral protein is not available, it is possible to characterize the structure of the protein using computational methods such as comparative modeling, also known as template-based or homology modeling. Comparative modeling relies on the presence of a structural template, a homologous protein from a related virus. Recently, we and several other groups showed that an accurate single- and multi-template comparative modeling allowed one to expand the structural repertoire of the SARS-CoV-2 proteome, by modeling Nsp4, Nsp14, and E proteins, as well as two additional domains of Nsp3 protein [9] and S [13]. Despite this progress, nine proteins remain structurally unresolved, most noticeably M protein, that plays a critical role in the structural organization of the viral envelope [14]. While de novo models of these proteins have been proposed that do not rely on the template structures [15,16], the accuracy of these models is difficult to evaluate, prompting development of the more advanced, integrative, approaches that use experimental data to guide the modeling process [17,18].

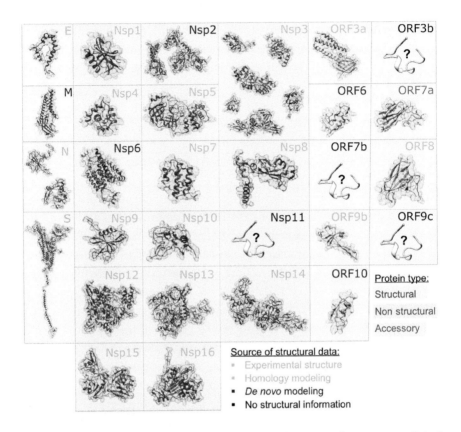

FIGURE 7.1 A "periodic" table of viral elements. Shown are three groups of viral proteins: (1) structural proteins (blue), nonstructural (red), and accessory (green). Presented in the table are either the entire structure of the protein experimentally resolved or modeled or individually resolved structural domains, as in the case of N and Nsp3 proteins. In addition, if a model uses both experimental structures and a model, it is denoted as a model overall (e.g., S protein).

7.4 STRUCTURAL CHARACTERIZATION OF INTRA-VIRAL AND VIRAL-HOST PROTEIN COMPLEXES

To perform a function, the majority of the viral proteins of SARS-CoV-2 must form a complex with the copies of itself, other viral proteins, host proteins, or RNA molecule. Understanding the structure of the macromolecular complexes is therefore important in elucidating the molecular function. Accurate modeling of protein complexes consisting of two or more subunits presents a significant challenge due to a substantially more limited number of homologous complexes to use in comparative modeling and low accuracy of models obtained using ab initio methods, such as protein docking [19]. In some cases, additional information such as complex symmetry or another experimental data can be used to further improve the model accuracy [20,21]. In other cases, multiple templates can be used to create a more complex model with the higher number of contributing proteins [22].

In our work, we used multichain and multi-template comparative modeling approaches using MODELLER [23] to reconstruct protein complexes of three kinds: viral homo-oligomers, viral hetero-oligomers, and viral-host complexes. As a result, we obtained structural models, including multiple conformations, for 16 homo-oligomeric protein complexes, three hetero-oligomeric complexes, and eight virus–human complexes [9]. The range of protein subunits varied from two, as in the case of M^{pro} homodimer or host-virus interaction between domain 5 of PL^{pro} and human ubiquitin, to five, as in E homopentamer (Figure 7.2a) and 12, as in Nsp10 homododecamer. The obtained models of the protein interaction complexes can be used to develop novel antivirals targeting the protein interaction interfaces and disrupting the viral function [24].

The information on protein binding sites extracted from the viral protein complexes can be also used to elucidate the evolutionary mechanisms and their impact on the viral protein function. Specifically, one can study a possible interaction-disrupting effect of non-synonymous mutations occurring at or near the protein-protein interaction interface [25,26]. Similar approaches can be used to analyze mutations within the same viral species [27,28] or across related viruses from a viral family [9,29,30]. Furthermore, if a protein-protein interaction is structurally resolved or modeled from a close homolog (Figure 7.2b), it is possible to apply more accurate methods that directly predict the rewiring, or edgetic, effect of a non-synonymous mutation [31]. Our analysis based on the structural comparison between homology models of the protein complexes and experimentally solved structures suggests that the modeling approach can be used as an accurate, fast, and inexpensive alternative to obtain the first crucial insights into molecular mechanisms behind the intraviral and viral-host interactions.

Recent studies have determined that regions of residues on viral protein surfaces that are conserved across different strains of a virus and even across evolutionary related viruses tend to colocalize with the intra-viral protein binding sites, while those regions consisting of rapidly mutating residues overlap with the protein binding sites involved in the viral-host protein-protein interactions [28–30]. First studies on the evolutionary diversity of coronaviruses and non-synonymous mutations in SARS-CoV-2 with high population incidence rates have provided several insights into the functional role of mutations [9,27]. In particular, four mutations in two proteins ORF3a and N were predicted to change the stability of the viral protein structure, binding affinity of potential intraviral interactions involving either of these proteins, and hot spots of the protein interaction interfaces.

In addition to studying genetic divergence across different lineages of SARS-CoV-2, one can investigate its evolutionary divergence by comparing with other related coronaviruses. In our work, SARS-CoV and two bat coronaviruses, BtCoV and BtRf-BetaCoV, were included in the comparative analysis with SARS-CoV-2. We found that all protein binding sites of nonstructural proteins that were involved in the intra-viral protein-protein interactions were either fully conserved or allowed at most one mutations on the periphery of the protein interaction interface, in spite of the fact that each of the interacting proteins had multiple mutations on its surface (Figure 7.2b). Similarly, a modeled host-viral protein-protein interaction between the papain-like protease PL^{pro} domain of Nsp3 and human ubiquitin-aldehyde (Figure 7.2c) revealed evolutionary conserved protein binding site on the PL^{pro} surface, with only two mutated residues located on the border of the binding region. In contrast, the protein binding surfaces of S responsible for interaction of the protein with the human ACE2 receptor and monoclonal antibodies appear to be heavily mutated. One of the questions that still remains is whether or not the mutated sites found when comparing related coronaviruses are the same as the ones found when comparing lineages of the same SARS-CoV-2 viral species. Finally, with a growing number of viral-host interaction complexes structurally resolved, such as the recently published macromolecular interaction between Nsp1 and human CCDC124-80S-eERF1 ribosome complex [32], the question of protein interaction specificity underlined by the genetic diversity of the homologous proteins can be now explored in substantially greater details.

7.5 MOLECULAR INTERACTIONS BETWEEN VIRAL PROTEINS AND SMALL LIGANDS

Determining the likely viral protein targets for therapeutic treatment with small molecular compounds is an important step in not only fighting the current SARS-CoV-2 virus, but in preparing for the future outbreaks of the related coronaviruses. Recently, several efforts, both experimental and computational, have been published that identify potential protein-ligand interactions between SARS-CoV-2 proteins and small ligands and

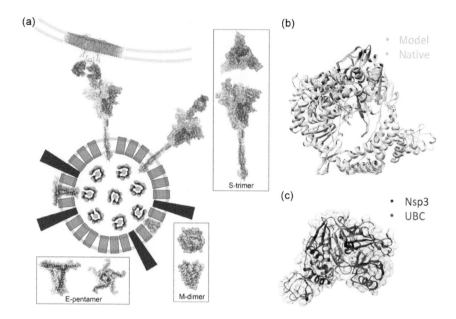

FIGURE 7.2 Structural modeling of intraviral and viral-host protein complexes. (a) Key components of the SARS-CoV-2 virion structure include an integrative model of S trimer shown interacting with human ACE2 receptor embedded into the membrane, an M dimer that constitutes the envelope wall, and an E pentamer whose structure is less studied. Shown inside the viral envelope are low-resolution models of RNP complexes (yellow) that form a viral capsid. (b) The multichain homology modeling of protein complexes allows to obtain accurate structures of large protein complexes months and sometimes years before they can be solved experimentally, presenting an important advancement in the structural characterization of a virus. The RMSD between a homology model of the protein complex and its experimental structure is only 1.55 Å. (c) Homology-based modeling of protein interactions can also provide a rapid structural assessment of host-virus complexes that have not been experimentally resolved for SARS-CoV-2.

characterize the structures of these interactions [7,33–36]. Nevertheless, the race for an efficient COVID-19 treatment is still on—only one drug candidate, remdesivir has been recently approved by FDA, but shows a somewhat modest improvement in the COVID-19 patients with moderate symptoms. Remdesivir is a broad-spectrum antiviral that was initially developed to target RNA polymerase in Ebola virus but is now repurposed to target a protein with the same function in SARS-CoV-2 [37].

Structural bioinformatics, and specifically computational methods for structure-based drug design, has become a critical first step in rapid identification of the new viral targets as well as finding new antivirals and repurposing existing drugs. In a recent work, a newly determined viral-host protein-protein interaction (PPI) network between the SARS-CoV-2 and human host factors was used as a basis to find potential antiviral candidates disrupting a host-viral PPI by binding the host factor [8]. In particular, the work identified 63 protein targets and 69 ligands that were either approved drugs, investigational drugs in clinical trials, or drugs in preclinical studies. The computational screening using cheminformatics analysis and molecular docking identified two previously approved drugs, Cloperastine and Clemastine, targeting human sigma-1 receptor to carry antiviral activity, whereas another, antitussive, compound Dextromethorphan exhibits proviral activity and therefore should be prescribed with caution. In several other recent studies, scientists used molecular docking to target SARS-CoV-2 proteins [7,34–36]. One of the key targets is the main protease, M^{pro}, that carries out an important function of digesting the polyprotein at multiple conserved sites [36]. The protease does not have any close homologs in human proteome, reducing potential side effects due to off-target binding and making it an attractive target for antivirals. It has been recently shown that an accurate homology modeling of the SARS-CoV-2 M^{pro} combined with molecular docking of a Michael acceptor inhibitor, N3, and followed by kinetic analysis could provide critical insights into the inhibition potency and the substrate-binding mode of N3 before the time-consuming and expensive experimental validation by X-ray crystallography was carried out. The same protease has also been the target of an ultrahigh-throughput screening using a deep docking approach [38]. Other important viral targets probed by the structure-based drug design approaches include PL^{pro}, RdRp, helicase, as well as structural proteins, S and E [33].

With more information available about the interactions between the viral proteins and their ligands, it is possible to use computational data-driven methods to address another question: how can the evolutionary changes accumulated in the viral genome within SARS-CoV-2 population or across known beta-coronaviruses affect binding of the viral inhibitors. The first step toward answering this question is to map the viral mutations onto the surface of a protein and estimate the proximity of these mutations to the ligand-binding sites. Two different scenarios could be considered with respect to the information available about the viral target and a ligand that binds it. In the first scenario, the information about structurally resolved viral-ligand complex would be available, allowing one to extract the ligand-binding site and to study the effects of viral mutations by mapping those mutations directly onto the protein surface and studying their colocalization with the functional site (Figure 7.3a). For instance, by combining multiple protein-ligand structures between the main protease M^{pro} and its novel inhibitors [36,39,40], one could determine that they target the same functional site of M^{pro}. Furthermore, by determining and mapping the evolutionary diverse and conserved regions, one can conclude that the functional site of M^{pro} is conserved across multiple viral species and presents an attractive drug target that will potentially be intact in the future viral strains (Figure 7.3a). In the second scenario, we do not have any structurally resolved interaction between the target viral protein and its ligand, while the structure of the protein may or may not be able available. However, there is structural information about a homolog of this protein interacting with a small ligand. In this case, one can first obtain a comparative model of the target protein using the structure of its homolog as a template, and then employ a structural alignment between the comparative model and the structural template to map the template's ligand-binding site on the surface of the comparative model. In our recent work, we have provided such mapping for the majority of structurally resolved proteins of SARS-CoV-2 using structural template proteins from closely related coronaviruses and their interactions with the small ligands [9]. Interestingly, when mapping sequence variations between SARS-CoV-2, SARS-CoV, and two closely related bat coronaviruses, we found that the majority of ligand binding sites remained intact. A question remains unanswered of whether mutations arising throughout the evolution of SARS-CoV-2 and through its geographic spread across different nations may affect the binding of the antiviral candidates. A similar question about a functional role of population-specific genetic variations can be asked when considering antivirals that inhibit the human receptor activity, potentially preventing or substantially weakening host-virus protein-protein interactions (Figure 7.3b) [41,42].

7.6 VIRUS-HOST INTERACTIONS: A SYSTEMS VIEW

To facilitate the understanding of SARS-CoV-2 infection and replication in a human host cell, it is important to breakdown this intracellular process into a set of mechanisms. This is possible by mapping the intracellular protein-protein interactions between the viral and host proteins. In our recent work [9], based on the genomic similarity between SARS-CoV-2 and SARS-CoV, the SARS-CoV-2 interactome was predicted using existing

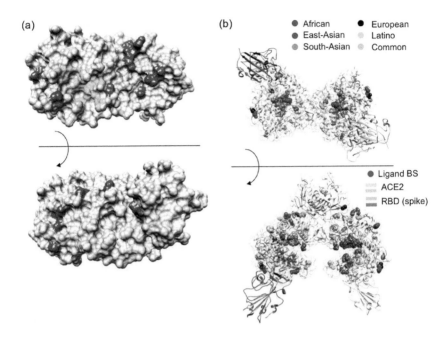

FIGURE 7.3 Integrating ligand binding site information and the genetic variation for SARS-CoV-2 and its human host to evaluate new antiviral therapies. (a) Evolutionary changes indicated by mapped protein mutations (orange) across 22 closely related coronaviruses mapped on the structure of main protease M^{pro} do not affect the functional site (blue) that is targeted by several new antivirals. (b) Similarly, a previously developed inhibitor of human ACE2 receptor (pink and blue) that could be potentially used to weaken or disrupt the interaction between the receptor and receptor-binding domain (RBD, light grey and dark grey) of S protein is unlikely to be affected by the population-specific mutations, since none of them are located at the corresponding ligand-binding site (orange).

data from SARS-CoV interactions. However, since then experimental studies have been published in which both the intraviral and virus-host PPIs have been characterized. Here, we created an extensive interactome by integrating data from the recent experiments and included annotation on the cellular components and biological processes potentially impacted by the viral proteins. With the help of this information, it could be possible to study the affected host proteins that influence downstream biological processes [43], providing new insights into drug treatment through the existing host targets and enabling the discovery of the new ones [44].

Here, we use two recent peer-reviewed SARS-CoV-2 experimental studies [8,45] that characterize many PPIs and integrate the individual

PPI networks to create an expanded network. We select these two specific studies because they both use the same affinity purification/mass spectrometry (AP-MS) approach to quantify the virus-host PPIs and have similar study designs. In the first work, 332 virus-host PPIs were discovered; however, no interactions were found that were mediated by ORF7b and NSP3 proteins due to method limitations [8]. Thus, we added another complimentary study—it is more recent and includes data on NSP3 interactions [45]. Furthermore, in the latter study, yeast two-hybrid screens and co-immunoprecipitation experiments were also performed for the genome-wide characterization of intraviral protein interactions.

To provide the information on the impacted cellular machinery and processes, we perform annotation of the host proteins using the Gene Ontology (GO) domains and terms [46,47]. For the annotation we used clusterProfiler [48] to obtain cellular component and biological process terms. The redundant GO terms were filtered out using the "simplify" functionality of clusterProfiler and were further manually grouped based on similarity. The interaction and annotation data were imported in Cytoscape [49] for visualization and network analysis. For the annotated proteins, we surveyed recent peer-reviewed literature to provide evidence on the host proteins, targeted by SARS-CoV-2 for its life cycle, that are parts of the intracellular machinery and pathways. By integrating the two studies on virus-host interactions, we found that only 15% of the interactions (45 PPIs) were common to both; therefore, the curated interactome had largely complementary information from both studies, resulting in 609 interactions between 28 viral proteins and 572 host proteins (Table 7.1). The resulting intraviral interactome also includes

TABLE 7.1   Network Characteristics of the SARS-CoV-2 Intraviral and Virus-Host Interactome

Network Statistics	Intraviral Interactome	Virus-Host Interactome
N of nodes	24	600
N of edges	57	609
N of components	1	12
Diameter	5	12
Average degree	4.75	2.03
Clustering coefficient	0.256	0.0

Shown are the topological characteristics for the two networks, including the number of nodes and edges in the networks, the average degree, number of components (independent sub-networks), diameter (maximum shortest path), and clustering coefficient (node degree at which clusters form).

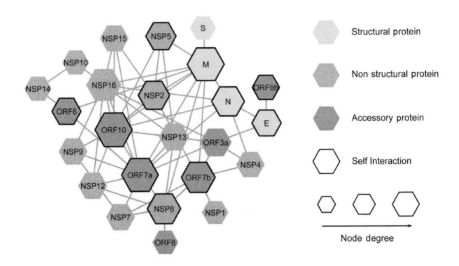

FIGURE 7.4   SARS-CoV-2 intraviral interactome. The interactome shows protein-protein interactions that were experimentally resolved through yeast two-hybrid screens and co-immunoprecipitation. There are 24 proteins and 57 interactions in total. The size of a node is proportional to the number of interactions, and dark outlines represent self-interactions.

self-interactions and highly interacting structural proteins that point to homo- and hetero-oligomers that perform functions in viral replication, possibly altering the host immunity (Figure 7.4).

We next provided the functional GO annotation of the host proteins that are involved in the virus-host interactome (Figure 7.5). The GO anno-tation domains consist of cellular component (CC) and biological process (BP). Since SARS-CoV-2 is known to alter host immune response [50,51], we specifically included the BP terms associated with the intracellular immune processes. The viral proteins of SARS-CoV and SARS-CoV-2 are known to hijack the cellular components responsible for protein trans-port: endoplasmic reticulum (ER), ER–Golgi intermediate compartment (ERGIC), and Golgi stack are primarily targeted by ORF3a and ORF8 in the current interactome [52,53]. Another important cell component is stress granule (SG), which is an aggregation of RNA molecules stalled dur-ing translation due to unfavorable conditions. The nucleocapsid protein (N) is shown to interact with at least seven host proteins that are part of stress granules (SGs). Recent studies regarding SARS-CoV, MERS-CoV, and SARS-CoV-2 have shown that coronaviruses can modify host cell translation that results in enhancing or inhibiting SG formation but helps

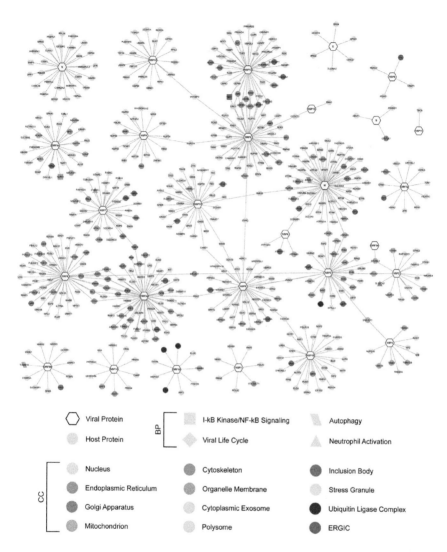

FIGURE 7.5 SARS-CoV-2 virus-host interactome. The interactome shows the aggregated protein-protein interactions from two complementary AP-MS studies resulting in 600 proteins and 609 interactions. The GO annotation for the host proteins shows the various cellular components (CC) and biological processes (BP) that are impacted. Specifically, the intracellular immune process is highlighted to show important pathways like autophagy and I-kappa B kinase/NF-kappa B signaling that are affected.

in viral replication [54,55]. Additionally, there are PPIs that are responsible for the formation of pathogen inclusion bodies [56].

The mitochondrial and organelle membrane proteins are targeted mainly by Nsp2, 3, 4, and ORF9c. The Type-I interferon (IFN-I) signaling pathway is essential to innate immunity and ORF9b is known [57] to inhibit IFN-I response by interacting with TOMM70 expressed on the mitochondrial outer membrane, and this interaction can also trigger autophagy. The viral proteins Nsp13, ORF6, and ORF9c interact with host proteins that are part of the I-kappa B kinase/NF-kappa B signaling pathway that plays a central part in the inflammasome and can modulate this process as it is responsible for the production of cytokines, inflammation, and apoptosis [58,59].

In addition to the comprehensive network created in our work, there have been two other recent studies on virus-host interactions. In one study, the authors use AP-MS and other high-throughput -omics analyses to characterize virus-host PPIs for both SARS-CoV and SARS-CoV-2, identifying the common host proteins targeted by the viruses [60]. In the other study, the authors identified a large number of membrane-associated PPIs using proximity-dependent biotin labeling (BioID), which can characterize transient and weak interactions, unlike AP-MS [61].

With more studies, our knowledge of the interactions between viral proteins and host proteins is expected to grow, gradually increasing the size of the virus-host interactome to provide a more complete picture and shedding light on the many ways this novel virus infects a host, replicates, and transmits. This will allow the identification of novel therapeutic targets aiming at disrupting key macromolecular interactions [62].

7.7 NEXT STEPS

In spite of the advancements in our understanding of the mechanistic nature of the virus, many things remain to be done. First, the structures of some of the key viral proteins alone and in their oligomeric states, as intra-viral complexes, have not been characterized yet, either experimentally or computationally, due to the lack of homologous templates, difficulties in some critical stages of experiments such as solubilization, purification, and crystallization and the complexity of the protein oligomeric states. In particular, we have yet to decipher the structures of M proteins, their basic dimeric complexes, and a higher order structural organization in the viral envelope. For another key structural protein, N, its multi-domain single-chain protein structure as well its supramolecular formation in a complex with viral RNA remains a mystery. Both proteins are the key components that constitute the molecular structure of a virion particle and are thus required in order to provide an accurate model of SARS-CoV-2 virion. Providing atomic details about the mechanistic process of attachment of S protein, specifically its receptor-binding domain (RBD), to human ACE2 receptor is another important question that could be addressed with structural biology and/or bioinformatics methods. For instance, it has been recently shown that a mutation D614G decreases the binding affinity for ACE2 through increasing the dissociation rate [63], and it is hypothesized that this change, in fact, would be reflected in mechanistic difference of receptor binding. The role of viral and human mutations in the vast viral-host PPI network is yet another topic that needs to be addressed in order to understand the future viral evolutionary trajectories and their functional impact on different human populations. In spite of the immense danger the current pandemics brings in, it also provides the scientific community with the unprecedented opportunity to study and understand the virus in great detail to contain the current and prevent future infections.

REFERENCES

1. Worldometers. COVID-19 Coronavirus Pandemic 2020. Available from: https://www.worldometers.info/coronavirus/.
2. Lu R, Zhao X, Li J, Niu P, Yang B, Wu H, et al. Genomic characterisation and epidemiology of 2019 novel coronavirus: Implications for virus origins and receptor binding. *The Lancet*. 2020;395(10224):565–74.
3. Ge X-Y, Li J-L, Yang X-L, Chmura AA, Zhu G, Epstein JH, et al. Isolation and characterization of a bat SARS-like coronavirus that uses the ACE2 receptor. *Nature*. 2013;503(7477):535–8.
4. de Groot RJ, Baker SC, Baric RS, Brown CS, Drosten C, Enjuanes L, et al. Commentary: Middle east respiratory syndrome coronavirus (mers-cov): Announcement of the coronavirus study group. *Journal of Virology*. 2013;87(14):7790–2.
5. Woo PC, Lau SK, Chu C-M, Chan K-H, Tsoi H-W, Huang Y, et al. Characterization and complete genome sequence of a novel coronavirus, coronavirus HKU1, from patients with pneumonia. *Journal of Virology*. 2005;79(2):884–95.
6. Gordon DE, Hiatt J, Bouhaddou M, Rezelj VV, Ulferts S, Braberg H, et al. Comparative host-coronavirus protein interaction networks reveal pan-viral disease mechanisms. *Science*. 2020;370(6521):eabe9403.
7. Francés-Monerris A, Hognon C, Miclot T, García-Iriepa C, Iriepa I, Terenzi A, et al. Molecular basis of SARS-CoV-2 infection and rational design of potential antiviral agents: Modeling and simulation approaches. *Journal of Proteome Research*. 2020;19(11):4291–4315.

8. Gordon DE, Jang GM, Bouhaddou M, Xu J, Obernier K, White KM, et al. A SARS-CoV-2 protein interaction map reveals targets for drug repurposing. *Nature*. 2020;583:459–68.

9. Srinivasan S, Cui H, Gao Z, Liu M, Lu S, Mkandawire W, et al. Structural genomics of SARS-CoV-2 indicates evolutionary conserved functional regions of viral proteins. *Viruses*. 2020;12(4):360.

10. Tang X, Wu C, Li X, Song Y, Yao X, Wu X, et al. On the origin and continuing evolution of SARS-CoV-2. *National Science Review*. 2020;7(6):1012–23.

11. Yao H, Song Y, Chen Y, Wu N, Xu J, Sun C, et al. Molecular architecture of the SARS-CoV-2 virus. *Cell*. 2020;183(3):730–8. e13.

12. Sedova M, Jaroszewski L, Alisoltani A, Godzik A. Coronavirus3D: 3D structural visualization of COVID-19 genomic divergence. *Bioinformatics*. 2020;36(15):4360–62.

13. Woo H, Park S-J, Choi YK, Park T, Tanveer M, Cao Y, et al. Modeling and simulation of a fully-glycosylated full-length SARS-CoV-2 spike protein in a viral membrane. *BioRxiv*.

14. Neuman BW, Kiss G, Kunding AH, Bhella D, Baksh MF, Connelly S, et al. A structural analysis of M protein in coronavirus assembly and morphology. *Journal of Structural Biology*. 2011;174(1):11–22.

15. Heo L, Feig M. Modeling of Severe Acute Respiratory Syndrome Coronavirus 2 (SARS-CoV-2) proteins by machine learning and physics-based refinement. *BioRxiv*. 2020.

16. Senior AW, Evans R, Jumper J, Kirkpatrick J, Sifre L, Green T, et al. Improved protein structure prediction using potentials from deep learning. *Nature*. 2020;577(7792):706–10.

17. MacCallum JL, Perez A, Dill KA. Determining protein structures by combining semireliable data with atomistic physical models by Bayesian inference. *Proceedings of the National Academy of Sciences*. 2015;112(22):6985–90.

18. Webb B, Viswanath S, Bonomi M, Pellarin R, Greenberg CH, Saltzberg D, et al. Integrative structure modeling with the integrative modeling platform. *Protein Science*. 2018;27(1):245–58.

19. Desta IT, Porter KA, Xia B, Kozakov D, Vajda S. Performance and its limits in rigid body protein-protein docking. *Structure*. 2020;28(9):1071–81. e3.

20. Yan Y, Tao H, Huang S-Y. HSYMDOCK: A docking web server for predicting the structure of protein homo-oligomers with Cn or Dn symmetry. *Nucleic Acids Research*. 2018;46(W1):W423–W31.

21. Xue LC, Rodrigues JP, Dobbs D, Honavar V, Bonvin AM. Template-based protein–protein docking exploiting pairwise interfacial residue restraints. *Briefings in Bioinformatics*. 2017;18(3):458–66.

22. Davis FP, Braberg H, Shen M-Y, Pieper U, Sali A, Madhusudhan M. Protein complex compositions predicted by structural similarity. *Nucleic Acids Research*. 2006;34(10):2943–52.

23. Webb B, Sali A. Comparative protein structure modeling using MODELLER. *Current Protocols in Bioinformatics*. 2016;54(1):5.6. 1–5.6. 37.

24. Schormann N, Sommers CI, Prichard MN, Keith KA, Noah JW, Nuth M, et al. Identification of protein-protein interaction inhibitors targeting vaccinia virus processivity factor for development of antiviral agents. *Antimicrobial Agents and Chemotherapy*. 2011;55(11):5054–62.

25. Goodacre N, Devkota P, Bae E, Wuchty S, Uetz P. Protein-protein interactions of human viruses. *Seminars in Cell and Developmental Biology*. 2020;99:31–9. Elsevier.

26. Brito AF, Pinney JW. Protein–protein interactions in virus–host systems. *Frontiers in Microbiology*. 2017;8:1557.

27. Wu S, Tian C, Liu P, Guo D, Zheng W, Huang X, et al. Effects of SARS-CoV-2 mutations on protein structures and intraviral protein-protein interactions. *Journal of Medical Virology*. 2020.

28. Warren S, Wan X-F, Conant G, Korkin D. Extreme evolutionary conservation of functionally important regions in H1N1 influenza proteome. *PLoS One*. 2013;8(11):e81027.

29. Neverov AD, Kryazhimskiy S, Plotkin JB, Bazykin GA. Coordinated evolution of influenza A surface proteins. *PLoS Genetics*. 2015;11(8):e1005404.

30. Voitenko OS, Dhroso A, Feldmann A, Korkin D, Kalinina OV. Patterns of amino acid conservation in human and animal immunodeficiency viruses. *Bioinformatics*. 2016;32(17):i685–i92.

31. Zhao N, Han JG, Shyu C-R, Korkin D. Determining effects of non-synonymous SNPs on protein-protein interactions using supervised and semi-supervised learning. *PLoS Computational Biology*. 2014;10(5):e1003592.

32. Thoms M, Buschauer R, Ameismeier M, Koepke L, Denk T, Hirschenberger M, et al. Structural basis for translational shutdown and immune evasion by the Nsp1 protein of SARS-CoV-2. *BioRxiv*. 2020.

33. Wu C, Liu Y, Yang Y, Zhang P, Zhong W, Wang Y, et al. Analysis of therapeutic targets for SARS-CoV-2 and discovery of potential drugs by computational methods. *Acta Pharmaceutica Sinica B*. 2020;10(5):766–8.

34. Li Z, Li X, Huang Y-Y, Wu Y, Liu R, Zhou L, et al. Identify potent SARS-CoV-2 main protease inhibitors via accelerated free energy perturbation-based virtual screening of existing drugs. *Proceedings of the National Academy of Sciences*. 2020;117(44):27381–7.

35. Zhou QA, Kato-Weinstein J, Li Y, Deng Y, Granet R, Garner L, et al. Potential therapeutic agents and associated bioassay data for COVID-19 and related human coronavirus infections. *ACS Pharmacology and Translational Science*. 2020;3(5):813–34.

36. Jin Z, Du X, Xu Y, Deng Y, Liu M, Zhao Y, et al. Structure of M pro from SARS-CoV-2 and discovery of its inhibitors. *Nature*. 2020;582:289–93.

37. Beigel JH, Tomashek KM, Dodd LE, Mehta AK, Zingman BS, Kalil AC, et al. Remdesivir for the treatment of Covid-19. *New England Journal of Medicine*. 2020;383:1813–26.

38. Ton AT, Gentile F, Hsing M, Ban F, Cherkasov A. Rapid identification of potential inhibitors of SARS-CoV-2 main protease by deep docking of 1.3 billion compounds. *Molecular Informatics*. 2020;39(8):e2000028.

39. Su H-X, Yao S, Zhao W-F, Li M-J, Liu J, Shang W-J, et al. Anti-SARS-CoV-2 activities in vitro of Shuanghuanglian preparations and bioactive ingredients. *Acta Pharmacologica Sinica*. 2020;41(9):1167–77.

40. Dai W, Zhang B, Jiang X-M, Su H, Li J, Zhao Y, et al. Structure-based design of antiviral drug candidates targeting the SARS-CoV-2 main protease. *Science*. 2020;368(6497):1331–5.

41. Towler P, Staker B, Prasad SG, Menon S, Tang J, Parsons T, et al. ACE2 X-ray structures reveal a large hinge-bending motion important for inhibitor binding and catalysis. *Journal of Biological Chemistry*. 2004;279(17):17996–8007.

42. Xiu S, Dick A, Ju H, Mirzaie S, Abdi F, Cocklin S, et al. Inhibitors of SARS-CoV-2 entry: Current and future opportunities. *Journal of Medicinal Chemistry*. 2020;63(21):12256–74.

43. Bojkova D, Klann K, Koch B, Widera M, Krause D, Ciesek S, et al. Proteomics of SARS-CoV-2-infected host cells reveals therapy targets. *Nature*. 2020;583(7816):469–72.

44. Sadegh S, Matschinske J, Blumenthal DB, Galindez G, Kacprowski T, List M, et al. Exploring the SARS-CoV-2 virus-host-drug interactome for drug repurposing. *Natural Communication*. 2020;11:3518.

45. Li J, Guo M, Tian X, Wang X, Yang X, Wu P, et al. Virus-host interactome and proteomic survey reveal potential virulence factors influencing SARS-CoV-2 pathogenesis. *Med*. 2020;2(1):99–112.e7.

46. Consortium GO. The gene ontology resource: 20 years and still GOing strong. *Nucleic Acids Research*. 2019;47(D1):D330–D8.

47. Ashburner M, Ball CA, Blake JA, Botstein D, Butler H, Cherry JM, et al. Gene ontology: Tool for the unification of biology. *Nature Genetics*. 2000;25(1):25–9.

48. Yu G, Wang L-G, Han Y, He Q-Y. clusterProfiler: An R package for comparing biological themes among gene clusters. *Omics: A Journal of Integrative Biology*. 2012;16(5):284–7.

49. Shannon P, Markiel A, Ozier O, Baliga NS, Wang JT, Ramage D, et al. Cytoscape: A software environment for integrated models of biomolecular interaction networks. *Genome Research*. 2003;13(11):2498–504.

50. Zohar T, Alter G. Dissecting antibody-mediated protection against SARS-CoV-2. *Nature Reviews Immunology*. 2020;20:392–4.

51. Blanco-Melo D, Nilsson-Payant BE, Liu W-C, Uhl S, Hoagland D, Møller R, et al. Imbalanced host response to SARS-CoV-2 drives development of COVID-19. *Cell*. 2020;181(5):1036–45.e9.

52. Amanat F, Krammer F. SARS-Co V-2 vaccines: Status report. *Immunity*. 2020;52(4):583–9.

53. Frieman M, Yount B, Heise M, Kopecky-Bromberg SA, Palese P, Baric RS. Severe acute respiratory syndrome coronavirus ORF6 antagonizes STAT1 function by sequestering nuclear import factors on the rough endoplasmic reticulum/Golgi membrane. *Journal of Virology*. 2007;81(18):9812–24.

54. Nakagawa K, Narayanan K, Wada M, Makino S. Inhibition of stress granule formation by Middle East respiratory syndrome coronavirus 4a accessory protein facilitates viral translation, leading to efficient virus replication. *Journal of Virology*. 2018;92(20):e00902.

55. Cascarina SM, Ross ED. A proposed role for the SARS-CoV-2 nucleocapsid protein in the formation and regulation of biomolecular condensates. *The FASEB Journal*. 2020;34(8):9832–42.

56. Morbini P, Benazzo M, Verga L, Pagella FG, Mojoli F, Bruno R, et al. Ultrastructural evidence of direct viral damage to the olfactory complex in patients testing positive for SARS-CoV-2. *JAMA Otolaryngology: Head & Neck Surgery*. 2020;146(10):972–3.

57. Jiang H-W, Zhang H-N, Meng Q-F, Xie J, Li Y, Chen H, et al. SARS-CoV-2 Orf9b suppresses type I interferon responses by targeting TOM70. *Cellular and Molecular Immunology*. 2020;17(9):998–1000.

58. Zbinden-Foncea H, Francaux M, Deldicque L, Hawley JA. Does high cardiorespiratory fitness confer some protection against pro-inflammatory responses after infection by SARS-CoV-2? *Obesity*. 2020;28(8):1378–81.

59. Fung S-Y, Yuen K-S, Ye Z-W, Chan C-P, Jin D-Y. A tug-of-war between severe acute respiratory syndrome coronavirus 2 and host antiviral defence: Lessons from other pathogenic viruses. *Emerging Microbes and Infections*. 2020;9(1):558–70.

60. Stukalov A, Girault V, Grass V, Bergant V, Karayel O, Urban C, et al. Multi-level proteomics reveals host-perturbation strategies of SARS-CoV-2 and SARS-CoV. *BioRxiv*. 2020.

61. St-Germain JR, Astori A, Samavarchi-Tehrani P, Abdouni H, Macwan V, Kim D-K, et al. A SARS-CoV-2 BioID-based virus-host membrane protein interactome and virus peptide compendium: new proteomics resources for COVID-19 research. *BioRxiv*. 2020.

62. Bobrowski T, Melo-Filho CC, Korn D, Alves VM, Popov KI, Auerbach S, et al. Learning from history: Do not flatten the curve of antiviral research! *Drug Discovery Today*. 2020;25(9):1604–13.

63. Yurkovetskiy L, Wang X, Pascal KE, Tomkins-Tinch C, Nyalile TP, Wang Y, et al. Structural and functional analysis of the D614G SARS-CoV-2 spike protein variant. *Cell*. 2020.

CHAPTER 8

Computational Tools for Discovery of CD8 T cell Epitopes and CTL Immune Escape in Viruses Causing Persistent Infections

Hadi Karimzadeh

University Hospital, Ludwig Maximilian University

Daniel Habermann and Daniel Hoffmann

University of Duisburg-Essen

Michael Roggendorf

Technical University of Munich

CONTENTS

The immune response of organisms to defend against a viral infection consists of two arms: the innate and adaptive immune system. The evolutionary older innate immune response acts immediately upon infection of cells and is independent of the amino acid sequence of the corresponding viral protein. Innate immunity can distinguish between cellular and viral proteins and nucleic acids in different classes of viruses by recognizing certain structural patterns resulting in an immediate induction of cytokines which downregulate the viral replication. This innate immune sensing often depends on characteristic modifications of viral genomes or DNA/RNA-replication intermediates and mRNA as well as special RNA structures which are normally absent in eukaryotic cells. The adaptive immune response by B cells results in antibody production, T helper cells, and CD8 T cells results in cytokine production and cytolytic activity (cytotoxic T lymphocytes, CTLs). This adaptive response needs hours or days to develop and depends on the amino acid sequences of viral proteins.

Peptides generated from viral proteins by proteasome cleavage are presented by the different MHC class I and II molecules on the cell surface and are recognized by T cell receptors in an amino acid sequence-specific manner.

8.1   IMPACT OF VIRAL MUTATIONS IN AMINO ACID SEQUENCE OF VIRAL PROTEINS ON EPITOPE RECOGNITION DURING CHRONIC INFECTION

In acute resolving viral infections, virus-specific CD8 T cells recognize epitopes of wildtype viral protein sequence presented by MHC class I on the surface of infected cells which results in a direct interaction with corresponding T cell receptors and subsequent cytokine production and cytotoxic activity by granzyme and perforin release. In persistent viral infections, mutations within viral protein occur during replication which can prevent various steps generation of peptides/viral epitopes leading to a productive T cell response, such as processing of viral peptides by proteasomes, binding with the corresponding MHC class I, or interaction with the T cell receptor. Prevention of any of these steps can lead to immune escape of the respective mutant virus. Characterization and prediction of epitopes present on a given amino acid sequence of a wild type virus are hampered by mutations that occurred during chronic infection or have occurred prior to transmission. Therefore, new methods have been established to define epitopes existing in wild-type virus which are absent in mutated viruses resulting in persistent infection.

In this review, new methods to determine CD8 T cell epitopes and immune escape mutations (IEMs) within virus causing chronic infections will be described. As a model system, to evaluate these new methods, the chronic infection with hepatitis delta virus (HDV) will be described as for this virus conventional prediction tools for epitope detection failed.

8.2 HDV AS A MODEL FOR DETECTION OF EPITOPES AND CORRESPONDING IMMUNE ESCAPE MUTATION IN CHRONIC VIRAL INFECTION

According to the World Health Organization (WHO), HDV affects an estimated 15–20 million out of the 240 million chronic hepatitis B virus (HBV) carriers [1]. New data published 2020 indicate that these numbers strongly underestimate the real prevalence of HDV [2].

Since HDV needs surface antigen from HBV for its envelope, HDV infection may occur as simultaneous infection together with HBV or as superinfection of a chronically HBV-infected patient. While simultaneous infection with HBV and HDV in adults mostly results in the resolution of both viruses, 80%–90% of patients with HDV superinfection develop chronic HBV/HDV infection. These patients are at great risk of fast progression to liver cirrhosis and liver failure. Therapeutic options are currently limited to pegylated interferon-alfa with low response rates. Vaccination strategies tested in preclinical models induced specific B- and T-cell responses and protected from simultaneous infection, however, failed to protect against superinfection of chronic HBV carrier animals.

In contrast to HBV, a strong innate immune response was observed in HDV infection [3]. These recent studies have shown that HDV leads to a robust activation of germline-encoded PRRs and antiviral molecules. This is of importance as it is a well-known fact that innate immunity shapes the adaptive immune system by context-dependent antigen presentation, expression of costimulatory molecules, and formation of a certain cytokine milieu affecting cells of adaptive immunity [4].

Patients who recovered from HDV infection (HDV RNA negative) show HDV-specific CD8 T cell responses. However, in the majority of patients with chronic HDV infection, virus specific T-cell response is absent. The number of identified epitopes is still very limited and epitopes may be primarily restricted by HLA-B alleles. IEMs within identified epitopes have been identified and occur frequently in the chronic stage of infection.

8.3 HDV MOLECULAR BIOLOGY AND REPLICATION

HDV virions particles are very small in size (about 36 nm in diameter) [5]. The viral genome of HDV, first described back in 1986 [6–8], is a single-stranded, circular, negative sense RNA with a length of ~1680 nucleotides the smallest described to infect human [9]. Due to a broad base pairing within the RNA molecule [7,8,10], the genome appears as a double stranded, rod-like structure resembling infectious agents in plants (viroids/virusoids). Nevertheless, there are a few differences between the plant subviral agents and HDV; for example, viroids' genome could be several folds smaller in size than that of HDV, viroids do not code for any protein, and they do not need a helper virus. With respect to virusoids, some of them do need a helper virus which provides an RNA-dependent RNA polymerase (RdRp) for the replication process; whereas, in the case of HDV, the helper virus (HBV) only provides the envelope proteins (HBs Ag) and HDV is indeed able to replicate in the absence of its helper, HBV. Based on this viroid-like genome, a double rolling circle mechanism was suggested as a model for viral RNA replication [11]. During HDV replication, exclusively taking place in the nucleus, three distinct RNAs, which include the genome, the positive-stranded antigenome, and viral mRNA, are generated by host RNA polymerases.

The single open reading frame within the HDV genome is actively transcribed and results in the expression of two isoforms of HDAg. The small HDAg (S-HDAg) is composed of 195 amino acids, and the large HDAg (L HDAg) comprises 214 amino acids. Initially, only S-HDAg is expressed because a termination codon prevents protein translation of L-HDAg. The latter is produced after the stop-codon (UAG) within the antigenome is mutated into a tryptophan codon (UGG) by the cellular enzyme Adenosine Deaminase Acting on RNA (ADAR1).

8.4 HDV GENOME VARIABILITY OF A SEQUENCE

Genetic variability of HDV, like many other RNA viruses, is very complex. In addition to the quasispecies formation by RNA viruses, the ADAR-dependent RNA editing adds more complexity to this. Since the first description of HDV genome, the proposed HDV nomenclature was updated once from three genotypes [12] to eight major clades, HDV-1 to HDV-8 [13]. Very recently two major papers, including one from our group, have suggested further classification determining subtypes with distinct geographical distributions [14,15]. HDV-1 is described to be the most prevalent HDV type in the world including North America, central Europe, Africa, some parts of Asia, and the western pacific region [16–19]. HDV-2, initially found in Japan and Taiwan [20,21], is mostly prevalent in the far East and some parts of North Asia [22], and HDV-4 is the next most prevalent type in Asia, mainly in Taiwan and Okinawa island [23]. HDV-3 was shown to be the most divergent one among all types [24], found exclusively in South America [12]. HDV-5 to HDV-8 were initially described in patients with African origin [10,13]; however, HDV-8 was recently reported from some patients in Brazil, as well [25]. Several studies, including experimental infection in woodchucks [26], evaluated the association between HDV (HBV) genotypes and the severity of the disease. Some of these studies indicated that infection with HDV-1, the worldwide distributed type, can lead to a wide range of disease outcomes from a very mild to a severe form of fulminant hepatitis [27] implying the need for further classification of some existing subtypes. HDV-3 (along with HBV-F) was described to be related to a more severe acute infection than other types [28,29]. On the contrary, HDV-2 and -4 were reported to be associated with a milder disease than HDV-1 and -3 [27]. Ultimately, it would be important to understand how these different HDV genotypes have an impact on clinical outcomes and response to therapy.

8.5 HDV IMMUNOLOGY

Virus-specific B and T cell immune response is thought to be essential for the elimination of hepatotropic viral infections. It has been studied in detail, e.g., in HBV and hepatitis C virus (HCV) infection [30–32]. In these infections, a concerted action of the different components of adaptive immunity has been shown to be required for viral clearance: B cells produce neutralizing antibodies that inhibit viral spread, CD4+ T cells provide important help to B and CD8+ T cells, and CD8+ T cells serve as main effector cells that have direct cytolytic activity and produce antiviral cytokines for non-cytolytic virus control.

HDV-specific immunity, in contrast, has been studied in little detail to date [33–37]. Studies of HDV-specific immunity are hampered by several aspects of HDV infection. First, HDV infection is rare compared, e.g., to HBV and HCV infection, especially in countries with high research resources such as the US or the European Union, with, e.g., only 30 cases of HDV infections reported annually in Germany. Second, it remains so far elusive if adaptive immunity in HDV infection targets primarily L-/S-HDAg as the only protein encoded by the HDV genome, or rather HBsAg, an essential component of the HDV envelope encoded by the HBV genome. In addition, virus-specific immunity in simultaneous HBV/HDV infection of an HBV-naïve individual may strongly differ from virus-specific immunity in HDV super-infection of an HBV-positive host already affected by HBV-specific T cell exhaustion.

Regarding HDV-specific antibodies, Mario Rizzetto and coworkers have demonstrated as early as 1979 (2 years after their first description of HDV) that HDV-specific antibodies are detectable only transiently and at low titers during acute-resolving infection, but are detectable at higher titers during persistent infection [38]. It has thus been concluded that HDV-specific antibodies have no virus neutralizing capacity and are therefore not required for HDV clearance.

In contrast to HBV and HCV infection, in which neutralizing antibodies are produced, in HDV superinfection chronic carriers of HBV antibodies to both proteins of HDV, P24 and P27, do not neutralize the HDV particle and hence cannot be expected to prevent the spread of the virus to terminate the infection. The HDV P24/P27/RNA complex is covered by the envelope protein of HBV (HBsAg) and chronic carriers of HBV have no anti-HB antibodies. Therefore, classical vaccines, which induce neutralizing antibodies, cannot prevent HDV superinfection of chronic HBV carriers. On the other hand, it has been shown that vaccines that induce virus-specific T cells were able to prevent infection by suppression of replication, e.g., by cytokine secretion. In a second step, these virus-specific T cells are able to eliminate infected cells by their cytolytic activity and thus prevent the spread of the virus [39]. T cell vaccines cannot provide sterile immunity, because they do not prevent infection of target cells [40,41]. However, T cell vaccines may eliminate infected cells by their cytotoxic activity, and stop the spread of the virus at a very early phase of infection. In contrast to HBV and HCV, in HDV infection initially a very low number of epitopes and IEMs had been identified by conventional techniques, despite the clear expectation of the existence of both. Fortunately, over the last years, there has been considerable progress in the development of bioinformatics methods that allow for the discovery of epitopes and IEMs from sequence data (paired with HLA information), as described below.

For virologists, these tools provide a new and efficient approach to a better understanding of key viral interactions with CTL immunity.

8.6 METHODS FOR THE PREDICTION OF CTL EPITOPES AND THE DETECTION OF IEMs

Presentation of peptides cut from viral proteins in MHC I molecules, and sensing of these peptide-MHC I complexes by CTLs is a key part of adaptive T cell immunity. Obviously, prediction of the presented peptides (=epitopes) is of great use for our understanding of immune responses to viral infections by HIV, HBV, HCV, HDV, Herpes viruses, and others, and for the design of vaccines. Therefore, there have been efforts to develop computational methods allowing such predictions, specific to given HLA alleles and amino acid sequences of antigens. These efforts have produced a series of tools of increasing predictive power, from earlier generations such as SYFPEITHI [42] to the current generation, with netMHCpan [43], MHCflurry [44], MixMHCpred [45], and other tools of related function; for a comparative review see Ref. [46]. Many of these programs focus on the prediction of MHC I binding peptides for given HLA alleles and protein sequence, because this is probably the tightest bottleneck that has to be overcome to initiate a CTL response.

The principle of these computer programs is generally the following: First, large datasets, resulting from MHC–peptide binding experiments, are compiled in a database (see, e.g., IEDB [47]). Second, a computational model, often an artificial neural network, is trained on the experimental data to predict those peptides that bind a given MHC molecule. Third, the users of the respective tool employ the model to predict, for a given amino acid sequence and HLA-allele, putative MHC binding peptides. These predictions usually come with scores for pairs of peptides and HLA-alleles, allowing the user to pick promising epitope candidates based on their better scores. The value of the models over a mere look-up list of measured peptide-affinity pairs is that these models extrapolate to affinities of unseen peptides.

Immune escape mutations (IEMs) are viral mutations that are selected by the immune system. In the case of T cell immunity, these mutations can interrupt the process of recognition by CTLs at several steps, including the complex processing of peptides in infected cells up to the presentation in MHC molecules, and the recognition of MHC-peptide complexes by T cell receptors. IEMs can render an otherwise presented peptide invisible to CTLs, giving the virus a selection advantage. IEMs often lie in epitopes and diminish or abolish the binding of the corresponding peptide to MHC. But they can also lie at sites that are important for peptide processing (proteasomal cleavage, trimming, MHC loading by TAP).

IEMs benefit the virus w.r.t. its visibility to the immune system and allow this population to expand within the host (Figure 8.1). On the other hand, IEMs can reduce the fitness of the virus by perturbing the function of the mutated viral protein. Therefore, IEMs are often accompanied by other viral mutations ("compensatory mutations") that help the virus to regain fitness. Since an unfit virus cannot really be called escaped, compensatory mutations also contribute to immune escape and therefore could be counted as IEMs in a general sense.

CTL IEMs will often be HLA-allele specific because MHC proteins, responsible for the binding of viral peptides, are encoded by the HLA-alleles. Therefore, IEMs can often be detected as HLA-allele-dependent footprints in sequence alignments of viral proteins: at certain alignment positions,

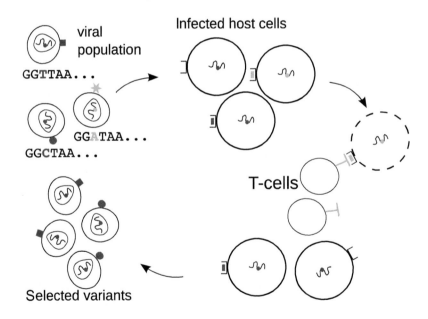

FIGURE 8.1 Viral in-patient evolution by immune selection: A population of genetically diverse virus particles (upper left) infects host cells. Viral antigen peptides may (green and blue) or may not (red) be presented by MHC I molecules on the cell surfaces. Cytotoxic T-cells detect viral epitopes (green) presented by MHC I molecules and destroy the cells (dashed cell membrane on right) infected by that viral variant. Cells infected by viral variants with immune escape mutations (red, green; bottom right) evade detection by T-cells and replicate (bottom left).

viral proteins coming from patients with a certain HLA-allele can have a different amino acid—an IEM, than the corresponding amino acids from patients of different HLA-alleles. Assuming for simplicity that we have only two amino acid states (X, Y) at a given alignment position, we can quantitatively detect IEMs for a specific HLA-allele H_i as follows. After the sequence alignment, we generate a 2-by-2 contingency table for each alignment position with the absolute frequency of number of sequences in the four categories (X, H_i), (X, H_*), (Y, H_i), (Y, H_*), where H_* stands for any HLA-allele except H_i. At the site of an IEM, we will find that sequences with H_i are dominated by one amino acid state, say X, while for other HLA-alleles H_* the other amino acid state prevails. Dominance or prevailing is typically not complete, and thus has to be quantified.

One way of quantification is Fisher's exact test (FET), as implemented in the SeqFeatR tool [48], which has been used increasingly in studies of IEMs. This includes a study with HDV, where the S170N polymorphism, located in an epitope predicted as having high affinity to HLA B15, was discovered with FET to be strongly associated with HLA B15 (Figure 8.2). FET is a conventional null hypothesis significance test (NHST) that generates p-values. The null hypothesis in this case is that HLA-allele and amino acid state are independent, and the p-value is the probability to find deviations from independence that are equal to or greater than the observed deviations from independence, assuming that the null hypothesis is true. In this application of FET, one of the test requirements is violated (fixed marginals), which in practice should usually not be problematic. SeqFeatR is fast and easy to use, but as NHST with p-values, FET suffers from problems that by now are widely appreciated in the literature [49], e.g., the nonsensical categorization with an arbitrary significance threshold and the potentially dire consequences of this procedure. Because of these shortcomings of NHST with p-values, the SeqFeatR software included also an alternative method, namely Bayesian inference based on the probability model of 2-by-2 contingency tables. There are other problems specific to NHST, such as the multiple testing problem, i.e., the control of false positives if many tests are performed (SeqFeatR sifts through a whole alignment and performs a test at every position), that are treated in SeqFeatR with standard corrections.

8.7 BETTER WAYS OF FINDING HLA-ASSOCIATED MUTATIONS (HAMs)

Strictly speaking, what we find in the way just outlined are HLA-associated mutations (HAMs). Given the underlying biology described above, HAMs

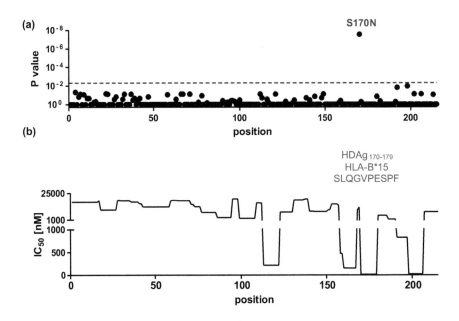

FIGURE 8.2 HLA-B*15-restricted CD8$_+$ T-cell responses specific for L-HDAg$_{170-179}$ drive viral escape. (Adapted from Ref. [35].) (a) P values for the association between HDV sequence polymorphisms and the presence of HLA-B*15, plotted for each amino acid residue in the L-HDAg protein. Cut-off for significance was set at $P = 0.005$ (dotted red line). (b) Predicted HLA-B*15 binding affinity of candidate peptide epitopes in L-HDAg. IC$_{50}$ values were predicted for 8mers, 9mers, and 10mers using the ANN 3.4 method (www.iedb.org). The best hit corresponding to the confirmed HLA-B*15-restricted epitope L-HDAg$_{170-179}$ is highlighted in green.

will often be IEMs. However, this is unfortunately not always the case. Apparent HAMs can occur for a number of other reasons, the most prominent of which is a phylogenetic bias: If in a part of the phylogenetic tree of viral sequences there happens to be an enrichment in certain mutations (perhaps due to a founder effect) *and* certain HLA-alleles, the above methods will discover an *apparent* HAM. SeqFeatR allows to estimate the strength of the phylogenetic bias for a candidate HAM with a simple approximation. It would be better to integrate estimates of the phylogenetic bias and other possible sources of apparent HAMs in a coherent model.

The first method offering such an integrated approach for the detection of HAMs that accounted for several of these sources was published in [50]. The model behind this method was much more complex than that of

SeqFeatR. It included confounding factors for HAM detection: (i) covariation of codons in the viral genome, (ii) linkage disequilibrium between HLA-alleles, and (iii) phylogenetic bias. First, covariation of codons refers mainly to the mentioned compensatory mutations. As there is often a physical, quasi-deterministic connection between first-line escape and compensatory mutations, both are effectively true HAMs, though the molecular mechanisms leading to their respective HLA-associations differ. Second, linkage disequilibrium is a more serious concern: some HLA-alleles frequently co-occur in the same person because they are genetically close and therefore often inherited together. In this case, there is usually a true association with one HLA-allele but only an apparent association with the other HLA-allele. A HAM detector should be able to distinguish these two possibilities. Third, phylogenetic bias is certainly a serious confounder that should be always addressed.

There are not only problems that confound HAM detection but there are also sources of information that can be used to detect HAMs more easily. For instance, IEMs are more likely to lie in or around CTL epitopes. Since we can predict CTL epitopes computationally to reasonable accuracy as described above, for given HLA-alleles and viral amino acid sequences, we have here a source of information that can impact the probability of a candidate HAM to be a true HAM. The same applies to other sources of information like, e.g., the prediction of cleavage sites of the immuno-proteasome. Bayesian methods allow for the inclusion of prior information in well-defined ways. We have therefore developed the Bayesian model HAM detector (manuscript in preparation) for inference of HAMs that can take advantage of information sources beyond HLA-sequence pairs, including phylogenetic relationships and predicted epitopes. Conceptually, HAM detector is a platform that can be extended to use further sources of information.

We applied a preliminary version of HAM detector to data from a recent publication on IEMs in HDV [35] (Table 8.1). The apparent HAMs in Table 8.1 had all p-values of <0.005 in a SeqFeatR analysis with FET. While there is no clear correlation of the p-values and experimental confirmation, the highest posterior probabilities computed with HAM detector are enriched in experimentally confirmed HAMs, and the unconfirmed HAM candidates have typically lower posterior probabilities. The fact that there are two false positives (B*14 with A107T, A*30 with D47E) shows that the model is a great improvement but likely not perfect.

TABLE 8.1 Results of HAM Detector Analysis of Confirmed or Predicted HAMs in HDV Sequences

Mutation	HLA	Post. Prob.	p-Value	Confirmed
S170N	B*15	0.99	2.9×10^{-8}	+
D101E	B*37	0.98	0.0002	+
P89T	B*37	0.97	0.0011	+
R139K	B*41	0.94	0.0034	+
A107T	B*14	0.93	0.0028	−
D47E	A*30	0.92	0.0010	−
R105K	B*27	0.91	0.0039	+
E47D	B*18	0.90	0.0027	+
D33E	B*13	0.86	0.0001	−
P49L	A*30	0.81	0.0031	−
K43R	B*13	0.76	0.0021	−
T134A	A*68	0.72	0.0045	−
K113R	B*13	0.68	0.0043	−
Q100K	B*13	0.56	0.0018	−
D96E	B*13	0.53	0.0035	−

Source: From Ref. [35].
For each pair of HDV mutation and HLA-allele (Column 1 and 2), the posterior probability (column3) is given. For comparison, the p-values obtained with SeqFeatR are presented. The last column indicates whether the HAM had been experimentally confirmed (+) or not (−).

8.8 CONCLUSION

There has been considerable progress over the last years in the development of bioinformatics methods to discover CD8 T cell epitopes and IEMs from viral amino acid sequence data. The benefits of the new bioinformatics methods are exemplified with the HDV infection in humans.

In the early times of mapping HLA epitopes by overlapping peptides and prediction tools of CD8 T-cell epitopes, understanding of immune response was extended and demonstrated for the first time that immune escape is a mechanism to establish persistent viral infections.

However initial studies, e.g., on CD8 T cell response in acute and chronic HDV infection surprisingly predicted only a very low number of epitopes that later on could be identified as being recognized by T cells (39). When we compared predicted CD8 T cell epitopes for the frequent HLA alleles of the nucleoprotein of HBV, HCV, and HDV, a low number was detected for HDV.

Application of the new bioinformatics tools to discover footprints described in this review allowed the identification of additional epitopes (Figure 8.3) of which at least five were recognized by CD8 T cells from

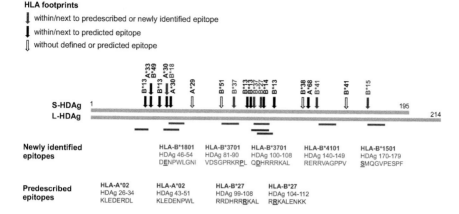

FIGURE 8.3 Viral sequence polymorphisms are associated with infrequent HLA class I alleles and spare the N- and C-termini of L-HDAg. (Adapted from Ref. [35].) Distribution of identified HLA footprints across L-HDAg. Red arrows indicate HLA footprints corresponding to confirmed (predescribed or newly identified) epitopes. Filled black arrows indicate HLA footprints corresponding to predicted epitopes. Empty black arrows indicate HLA footprints without defined or predicted epitopes. Newly identified epitopes are shown as red bars, and predescribed epitopes are shown as blue bars.

patients who recovered from infection. These new bioinformatics tools exemplified for HDV in this review are specifically of great value in chronic persistent viral infections, e.g., HBV, HCV, HIV, or Herpes viruses. These viruses allow adaptation to their host by creating IEMs which contribute to the persistence of infection due to ineffective CD8 T cells.

REFERENCES

1. Chen, H.Y., et al., Prevalence and burden of hepatitis D virus infection in the global population: A systematic review and meta-analysis. *Gut*, 2019. **68**(3): pp. 512–21.
2. Vlachogiannakos, J. and G.V. Papatheodoridis, New epidemiology of hepatitis delta. *Liver Int*, 2020. **40**(Suppl 1): pp. 48–53.
3. Jung, S., S.M. Altstetter, and U. Protzer, Innate immune recognition and modulation in hepatitis D virus infection. *World J Gastroenterol*, 2020. **26**(21): pp. 2781–91.
4. Jain, A. and C. Pasare, Innate control of adaptive immunity: Beyond the three-signal paradigm. *J Immunol*, 2017. **198**(10): pp. 3791–800.
5. Rizzetto, M., et al., delta Agent: Association of delta antigen with hepatitis B surface antigen and RNA in serum of delta-infected chimpanzees. *Proc Natl Acad Sci U S A*, 1980. **77**(10): pp. 6124–8.

6. Chen, P.J., et al., Structure and replication of the genome of the hepatitis delta virus. *Proc Natl Acad Sci U S A*, 1986. **83**(22): pp. 8774–8.

7. Kos, A., et al., The hepatitis delta (delta) virus possesses a circular RNA. *Nature*, 1986. **323**(6088): pp. 558–60.

8. Wang, K.S., et al., Structure, sequence and expression of the hepatitis delta (delta) viral genome. *Nature*, 1986. **323**(6088): pp. 508–14.

9. Rizzetto, M., The delta agent. *Hepatology*, 1983. **3**(5): pp. 729–37.

10. Radjef, N., et al., Molecular phylogenetic analyses indicate a wide and ancient radiation of African hepatitis delta virus, suggesting a deltavirus genus of at least seven major clades. *J Virol*, 2004. **78**(5): pp. 2537–44.

11. Branch, A.D. and H.D. Robertson, A replication cycle for viroids and other small infectious RNA's. *Science*, 1984. **223**(4635): pp. 450–5.

12. Casey, J.L., et al., A genotype of hepatitis D virus that occurs in northern South America. *Proc Natl Acad Sci U S A*, 1993. **90**(19): pp. 9016–20.

13. Le Gal, F., et al., Eighth major clade for hepatitis delta virus. *Emerg Infect Dis*, 2006. **12**(9): pp. 1447–50.

14. Karimzadeh, H., et al., Genetic diversity of hepatitis D virus genotype-1 in Europe allows classification into subtypes. *J Viral Hepat*, 2019. **26**(7): pp. 900–10.

15. Le Gal, F., et al., Genetic diversity and worldwide distribution of the deltavirus genus: A study of 2,152 clinical strains. *Hepatology*, 2017. **66**(6): pp. 1826–41.

16. Chao, Y.C., et al., Sequence conservation and divergence of hepatitis delta virus RNA. *Virology*, 1990. **178**(2): pp. 384–92.

17. Han, M., et al., Molecular epidemiology of hepatitis delta virus in the Western Pacific region. *J Clin Virol*, 2014. **61**(1): pp. 34–9.

18. Shakil, A.O., et al., Geographic distribution and genetic variability of hepatitis delta virus genotype I. *Virology*, 1997. **234**(1): pp. 160–7.

19. Zhang, Y.Y., E. Tsega, and B.G. Hansson, Phylogenetic analysis of hepatitis D viruses indicating a new genotype I subgroup among African isolates. *J Clin Microbiol*, 1996. **34**(12): pp. 3023–30.

20. Imazeki, F., M. Omata, and M. Ohto, Heterogeneity and evolution rates of delta virus RNA sequences. *J Virol*, 1990. **64**(11): pp. 5594–9.

21. Lee, C.M., et al., Characterization of a new genotype II hepatitis delta virus from Taiwan. *J Med Virol*, 1996. **49**(2): pp. 145–54.

22. Ivaniushina, V., et al., Hepatitis delta virus genotypes I and II cocirculate in an endemic area of Yakutia, Russia. *J Gen Virol*, 2001. **82**(Pt 11): pp. 2709–18.

23. Chang, S.Y., et al., Molecular epidemiology of hepatitis D virus infection among injecting drug users with and without human immunodeficiency virus infection in Taiwan. *J Clin Microbiol*, 2011. **49**(3): pp. 1083–9.

24. Deny, P., Hepatitis delta virus genetic variability: From genotypes I, II, III to eight major clades? *Curr Top Microbiol Immunol*, 2006. **307**: pp. 151–71.

25. Barros, L.M., et al., Hepatitis delta virus genotype 8 infection in Northeast Brazil: Inheritance from African slaves? *Virus Res*, 2011. **160**(1–2): pp. 333–9.

26. Parana, R., et al., Serial transmission of spongiocytic hepatitis to woodchucks (possible association with a specific delta strain). *J Hepatol*, 1995. **22**(4): pp. 468–73.

27. Wu, J.C., et al., Genotyping of hepatitis D virus by restriction-fragment length polymorphism and relation to outcome of hepatitis D. *Lancet*, 1995. **346**(8980): pp. 939–41.
28. Manock, S.R., et al., An outbreak of fulminant hepatitis delta in the Waorani, an indigenous people of the Amazon basin of Ecuador. *Am J Trop Med Hyg*, 2000. **63**(3–4): pp. 209–13.
29. Nakano, T., et al., Characterization of hepatitis D virus genotype III among Yucpa Indians in Venezuela. *J Gen Virol*, 2001. **82**(Pt 9): pp. 2183–9.
30. Thimme, R., et al., CD8(+) T cells mediate viral clearance and disease pathogenesis during acute hepatitis B virus infection. *J Virol*, 2003. **77**(1): pp. 68–76.
31. Timm, J. and C.M. Walker, Mutational escape of CD8+ T cell epitopes: Implications for prevention and therapy of persistent hepatitis virus infections. *Med Microbiol Immunol*, 2015. **204**(1): pp. 29–38.
32. Wieland, D., et al., TCF1+ hepatitis C virus-specific CD8+ T cells are maintained after cessation of chronic antigen stimulation. *Nat Commun*, 2017. **8**: p. 15050.
33. Huang, Y.H., et al., Identification of novel HLA-A*0201-restricted CD8+ T-cell epitopes on hepatitis delta virus. *J Gen Virol*, 2004. **85**(Pt 10): pp. 3089–98.
34. Karimzadeh, H., et al., Amino acid substitutions within HLA-B*27-restricted T cell epitopes prevent recognition by hepatitis delta virus-specific CD8(+) T cells. *J Virol*, 2018. **92**(13): pp. e01891-17.
35. Karimzadeh, H., et al., Mutations in hepatitis D virus allow it to escape detection by CD8(+) T cells and evolve at the population level. *Gastroenterology*, 2019. **156**(6): pp. 1820–33.
36. Kefalakes, H., et al., Hepatitis D virus-specific CD8(+) T cells have a memory-like phenotype associated with viral immune escape in patients with chronic hepatitis D virus infection. *Gastroenterology*, 2019. **156**(6): pp. 1805–19.e9.
37. Landahl, J., et al., Detection of a broad range of low-level major histocompatibility complex class II-restricted, Hepatitis Delta Virus (HDV)-specific T-cell responses regardless of clinical status. *J Infect Dis*, 2019. **219**(4): pp. 568–77.
38. Rizzetto, M., Hepatitis D: Thirty years after. *J Hepatol*, 2009. **50**(5): pp. 1043–50.
39. Fiedler, M., et al., Prime/boost immunization with DNA and adenoviral vectors protects from hepatitis D virus (HDV) infection after simultaneous infection with HDV and woodchuck hepatitis virus. *J Virol*, 2013. **87**(13): pp. 7708–16.
40. Fiedler, M. and M. Roggendorf, Immunology of HDV infection. *Curr Top Microbiol Immunol*, 2006. **307**: pp. 187–209.
41. Roggendorf, M., et al., Il Pensiero Scientifico Editore; Roma, Italy; New developments for prophylactic and therapeutic vaccines against HDV. In: Hepatitis D. *Virology, Management and Methodology, 2019*, pp.279–94.
42. Rammensee, H., et al., SYFPEITHI: Database for MHC ligands and peptide motifs. *Immunogenetics*, 1999. **50**(3–4): pp. 213–9.

43. Reynisson, B., et al., NetMHCpan-4.1 and NetMHCIIpan-4.0: Improved predictions of MHC antigen presentation by concurrent motif deconvolution and integration of MS MHC eluted ligand data. *Nucleic Acids Res*, 2020. **48**(W1): pp. W449–54.

44. O'Donnell, T.J., A. Rubinsteyn, and U. Laserson, MHCflurry 2.0: Improved pan-allele prediction of MHC class I-presented peptides by incorporating antigen processing. *Cell Syst*, 2020. **11**(1): pp. 42–48.e7.

45. Bassani-Sternberg, M., et al., Deciphering HLA-I motifs across HLA peptidomes improves neo-antigen predictions and identifies allostery regulating HLA specificity. *PLoS Comput Biol*, 2017. **13**(8): p. e1005725.

46. Mei, S., et al., A comprehensive review and performance evaluation of bioinformatics tools for HLA class I peptide-binding prediction. *Brief Bioinform*, 2020. **21**(4): pp. 1119–35.

47. Vita, R., et al., The Immune Epitope Database (IEDB): 2018 update. *Nucleic Acids Res*, 2019. **47**(D1): pp. D339–43.

48. Budeus, B., J. Timm, and D. Hoffmann, SeqFeatR for the discovery of feature-sequence associations. *PLoS One*, 2016. **11**(1): p. e0146409.

49. Amrhein, V., S. Greenland, and B. McShane, Scientists rise up against statistical significance. *Nature*, 2019. **567**(7748): pp. 305–7.

50. Carlson, J.M., et al., Phylogenetic dependency networks: Inferring patterns of CTL escape and codon covariation in HIV-1 Gag. *PLoS Comput Biol*, 2008. **4**(11): p. e1000225.

CHAPTER 9

Virus-Host Transcriptomics

Caroline C. Friedel

Ludwig-Maximilians-Universität München

CONTENTS

9.1 INTRODUCTION

Advances in next-generation sequencing (NGS) technologies now allow studying gene expression and transcription at genome-wide level with nucleotide resolution. While the most well-known technology for this purpose is RNA sequencing (RNA-seq) to measure total RNA abundances in a cell (Wang et al. 2009), the combination of RNA-seq with different approaches for extracting RNAs of interest provides even more detailed pictures of transcriptional processes. For instance, 4sU- and TT-seq use metabolic labeling of newly transcribed RNA using 4-thiouridine (4sU) to quantify de novo transcription and RNA processing and splicing (Windhager et al. 2012, Schwalb et al. 2016). Nuclear run-on sequencing (e.g. GRO- and PRO-seq) allows mapping the positions of actively transcribing RNA polymerases and thus ongoing nascent transcription (Mahat

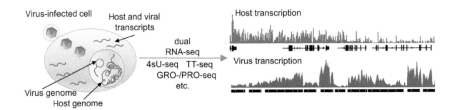

FIGURE 9.1 Schematic outline of dual RNA-seq for virus-host transcriptomics. Application of RNA-seq technologies to virus-infected cells yields sequencing reads from both host and viral RNAs. Depending on the sequencing protocol used, this allows quantification of both host and viral RNA levels (using total RNA-seq), de novo transcription (using 4sU-/TT-seq), ongoing nascent transcription (using GRO-/PRO-seq), and more.

et al. 2016, Core et al. 2008). For a review of other NGS-based technologies, see Soon et al. (2013).

In contrast to microarrays, RNA-seq does not require the prior definition of probes covering transcripts of interests. Thus, application to pathogen-infected cells allows easy profiling of both host and pathogen transcription in parallel using standard RNA extraction and library preparation protocols (Figure 9.1). Westermann et al. coined the term "dual RNA-seq" for this specific application (Westermann et al. 2012). The term has not really caught on, likely because it implies an inherent experimental difference to "standard" RNA-seq, which is not actually the case. For the sake of brevity, I will nevertheless use it in this chapter to denote some variant of RNA-seq, e.g., total RNA-seq, 4sU-/TT-seq, or GRO/-PRO-seq, performed on virus-infected cells.

Dual RNA-seq in its many variants has previously been applied to infections with several DNA and RNA viruses. This includes influenza (Cao et al. 2019, Bauer et al. 2018); several herpesviruses, in particular herpes simplex virus 1 (HSV-1) (Birkenheuer et al. 2018, Pheasant et al. 2018, Whisnant et al. 2020, Hennig et al. 2018, Rutkowski et al. 2015); and most recently SARS-CoV-2 (Blanco-Melo et al. 2020), to name just a few examples. In this chapter, I will outline the major challenges for bioinformatics analyses of dual RNA-seq data. The example I will most commonly employ for this purpose is herpesvirus infection, in particular HSV-1 infection, due to my extensive experience in virus-host transcriptomics of herpesviruses (Marcinowski et al. 2012, Rutkowski et al. 2015, Hennig et al. 2018, Wang et al. 2020, Whisnant et al. 2020, Friedel et al. 2021).

Herpesviruses are a family of DNA viruses with double-stranded DNA genomes (Pellett and Roizman 2013, Weidner-Glunde et al. 2020), eight of which are known to infect humans. Apart from HSV-1 and the closely related herpes simplex virus type 2 (HSV-2), this includes Epstein-Barr virus (EBV), which can cause mononucleosis but also a wide range of malignancies (Maeda et al. 2009), and varicella-zoster virus (VZV), the causative agent of chickenpox and shingles. While HSV-1 is widely known for causing mostly harmless cold sores on the lips, it can cause life-threatening diseases in both immunodeficient and immunocompetent individuals (Roizman et al. 2007). A defining feature of the herpesvirus life cycle is its two distinct phases: (i) a lytic phase with active replication of the virus and (ii) a latent phase with very limited viral gene expression and no production of viral particles (Weidner-Glunde et al. 2020). After initial lytic infection, herpesvirus infections enter the latent phase, which allows them to persist for the lifetime of the host, but they can be reactivated to lytic infection. Symptoms are only observed in the lytic phase; however, many infected hosts remain completely asymptomatic during both the initial and reactivated lytic phases and thus are unaware of their infection. Not surprisingly, studies on seroprevalence of herpesvirus antibodies commonly indicate that close to 100% of the human adult population are infected with one or more herpesviruses (see Wrensch et al. 2001). In this chapter, I will extensively refer to a 4sU-seq time-course experiment we performed with 1 hour 4sU labeling applied in hourly intervals of the first 8 hours of lytic HSV-1 infection (Rutkowski et al. 2015). In addition, we analyzed RNA-seq data of subcellular RNA fractions, i.e., cytosolic, nucleoplasmic, and chromatin-associated RNA, in mock infection and at 8 hours post infection (p.i.) (Hennig et al. 2018).

9.2 PARALLEL READ ALIGNMENT TO HOST AND VIRUS

The first major step of any NGS data analysis is the alignment of reads against a reference genome (Figure 9.2). This step is often denoted as read mapping. In case of RNA-seq and its variants, a splicing-aware read aligner such as STAR (Dobin et al. 2013) or HISAT2 (Kim et al. 2015) has to be used to align spliced reads that do not align contiguously to the genome. Alternatively, reads can be aligned to transcript sequences; however, this requires methods for assigning reads that can be ambiguously aligned to multiple transcripts as well as a well-annotated transcriptome. As a consequence, splicing-aware RNA-seq aligners generally only use the

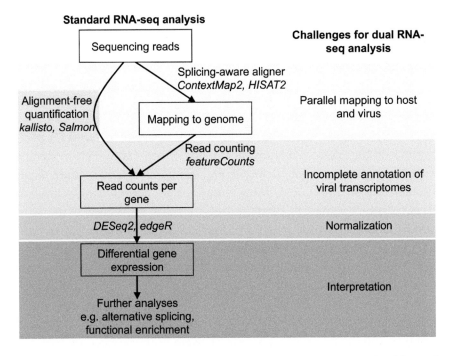

FIGURE 9.2 Overview of standard RNA-seq analysis steps and additional challenges arising in dual RNA-seq analysis. Example software packages for individual steps are indicated in italics.

transcriptome as additional, optional information to characterize exon boundaries. Transcriptome-only approaches are commonly employed by fast alignment-free methods for transcript quantification like kallisto (Bray et al. 2016) or Salmon (Patro et al. 2017). However, these methods do not produce nucleotide-level read alignments to specific locations in transcripts or the genome. Instead they produce so-called pseudo-alignments or quasi-mappings of reads to transcripts or groups of transcripts that allow quantifying transcript or gene expression.

Both alignment-based and alignment-free methods can essentially be applied "as-is" to dual RNA-seq data. The only additional challenge that arises is how to include the virus reference genome or transcriptome in the process. The most straightforward way is to add the virus genome as an additional chromosome to the host reference genome or, in case of alignment-free methods, add the virus transcripts to the host transcriptome. The disadvantage of this approach is that it requires creating new index files for the corresponding alignment or alignment-free program whenever a new virus is analyzed and potentially having to store multiple index files that always include the same large host genome as well as different small virus genomes.

An alternative approach performs read alignment separately against host and virus genomes, either in parallel or in sequence. However, this requires methods for resolving multiple alignments to the different genomes obtained in the parallel alignment runs or filtering unmapped or badly aligned reads for subsequent alignment steps.

We previously developed the RNA-seq mapper ContextMap2, which allows parallel mapping of reads against several reference sequences at the same time and choosing the best alignment for each read in any of these read sources (Bonfert et al. 2013, 2015). ContextMap2 evaluates multiple possible alignments for each read and chooses the best alignment not only based on the quality of the alignment but also based on alignments of other reads in the same region. For this purpose, initial alignments are clustered into so-called contexts, which generally cover one or a few genes. Alignments are then extended to include alternative spliced read alignments, in particular alignments crossing multiple introns. Subsequently, the best alignment for each read is first chosen within each context and then between contexts. There are two major advantages of ContextMap2 for dual RNA-seq analysis. First, it already implements a method for evaluating alternative alignments to different reference sequences and choosing the best one. Second, the implementation allows providing an arbitrary number of indices for the underlying alignment program. For instance, the host genome can be included as one index and one or more virus genomes as separate indices. Thus, application to a new virus only requires creating and storing one additional small index for this particular virus. We commonly also include rRNA sequences as a separate index as these are not well represented in the human reference genome. For some RNA-seq approaches, e.g., 4sU-seq, rRNA content can be significant as rRNA depletion is not practical due to the low RNA yield of separating the labeled RNA fraction. Similarly, poly(A) selection is not an option as many newly transcribed RNAs may not yet be polyadenylated.

It should be noted that the alignment to virus genomes may be omitted if only the transcriptional response of the host is of interest. In this case, previously developed analysis workflows for standard RNA-seq analysis can be used without modification. However, I would recommend against this approach for several reasons. First, one measure of data quality is the fraction of reads that can be aligned, which cannot be properly assessed if the virus genome is not included in the alignment. Second, both the fraction of reads aligned to the viral genome and expression of specific viral genes provide information with regard to the progression of the infection. This is of particular importance if one wants to compare results from different experiments as batch effects can also affect how fast infection progresses. Thus, at the same time-point of infection the actual stage of infection may differ between independent experiments. Third, local similarities between the host and viral genome may lead to misalignment of viral reads to the host genome, which can be avoided if the virus genome is included in the alignment. This is particularly the case if an RNA-seq mapper like ContextMap2 is used that also considers other reads aligned in the same general region to resolve ambiguous alignments. Local similarities can, e.g., originate from horizontal gene transfer from virus to host or vice versa (Blinov et al. 2017). Here, RNA retroviruses are the major source of viral DNA in the human genome, with endogenous retroviruses representing up to 8% of its total size. However, the incorporation of viral sequences into the human genome has also been reported for nonretroviral RNA and DNA viruses, e.g., Ebola, filo- and bornaviruses, dependoparvoviruses, and herpesviruses (see Blinov et al. 2017, for a review). Finally, viral genes are commonly densely packed on both strands of the genome and rarely spliced, resulting in almost pervasive transcription of the viral genome. As a consequence, RNA-seq reads can be used to extract the genome sequence of the virus and identify mutations compared to the reference genome. This is particularly useful when analyzing mutant viruses as the mutation can be automatically confirmed in the same experiment.

9.3 INCOMPLETE ANNOTATION OF VIRAL TRANSCRIPTOMES

Viral genes and proteins are often poorly characterized and annotated. This complicates bioinformatics analysis of dual RNA-seq data at every step from read alignment to differential gene or splicing analysis (Figure 9.2). Even after extensive studies over many decades, the true coding and noncoding potential of a virus can still be poorly understood. This is strikingly exemplified by the case of human cytomegalovirus (HCMV), another member of the herpesvirus family. The original catalog of HCMV genes included 189 protein-coding ORFs (Chee et al. 1990), which was later reduced to 147 genes (Davison et al. 2003) and subsequently again increased to 157 (Yu et al. 2003), 171 (Murphy, et al. 2003a), 220 (Murphy et al. 2003b) and 232 (Varnum et al. 2004) genes. A reason for this difficulty in defining viral genes is the often complex structure of viral transcriptomes with overlapping and antisense transcripts. This is particularly the case for herpesviruses with their relatively large genomes and large number of genes. Here, genome-wide functional genomics approaches, including not only NGS-based technologies but also quantitative proteomics, provide a possible solution to resolving the viral transcriptome and proteome. However, they may also lead to new confusion and uncertainty. For instance, recent large-scale RNA-seq and ribosome profiling studies indicated that the coding capacity of HCMV (Stern-Ginossar et al. 2012), EBV (Bencun et al. 2018), and a further human herpesvirus, Kaposi's sarcoma-associated herpesvirus (KSHV) (Arias et al. 2014), was much larger than previously suspected. Specifically, hundreds of viral gene products resulting from alternative transcription and translation start sites were identified for HCMV and KSHV in addition to hundreds of short ORFs (sORFs) of unknown function. Most of these sORFs have not been experimentally validated, and it remains unclear whether these are not artifacts from noise in the sequencing data.

The problem of incomplete viral gene catalogs for transcriptomic analyses is further exacerbated by the absence of a central database for up-to-date virus genome annotations. For vertebrates, Ensembl provides such a comprehensive repository with a wide range of access options, standardized formats, and regular updates (Yates et al. 2019). In contrast, the best general resource for virus gene annotations are genome sequences submitted to GenBank (Benson et al. 2013), which commonly include an initial gene annotation that may or may not have been updated based on more recent studies. Often enough updated gene annotations can only be found in one or more supplementary tables of more recent studies, which are not always in standard formats that can be easily used for bioinformatics analysis. Thus, precise viral gene expression profiling using dual RNA-seq data often requires some extent of data curation to obtain the most up-to-date annotation. The easier and more commonly used approach is to simply use the GenBank annotation; however in this case results—in particular surprising ones—should be taken with at least a modicum of skepticism.

9.4 NORMALIZATION AND DIFFERENTIAL GENE EXPRESSION ANALYSIS

Following read alignment and counting of aligned reads per gene, the next step in RNA-seq analysis is generally differential gene expression (DGE) analysis (Figure 9.2). A number of methods and software tools for DGE analysis have previously been developed specifically for gene read count data. Examples of the most commonly used methods are DESeq2 (Love et al. 2014) or edgeR (Robinson et al. 2009). While these methods are of course also applicable to host and/or viral gene read counts from dual RNA-seq data, they are generally based on the assumption that not all genes are differentially expressed. This assumption is likely violated in dual RNA-seq of virus-infected cells due to the fact that the additional expression of viral genes either increases overall RNA levels in the cells, globally reduces host gene expression, or both. For instance, HSV-1 globally downregulates host gene expression by recruiting RNA polymerase II and elongation factors from the host chromatin to the replicating viral genomes (Abrisch et al. 2016, Birkenheuer et al. 2018). In addition, several viruses, including HSV-1, KSHV and EBV (Kwong and Frenkel 1987, Glaunsinger and Ganem 2004, Rowe et al. 2007), influenza A virus (Jagger et al. 2012), and SARS coronavirus (Kamitani et al. 2009), encode viral proteins that trigger a widespread degradation of host mRNAs by inducing internal cleavages in these mRNAs.

One approach to normalize for these global changes in gene expression in virus-infected cells is to use RNA spike-ins, in particular those developed by the External RNA Controls Consortium (ERCC) (Jiang et al. 2011). The latter consists of 92 in vitro transcribed poly(A)+ RNAs with low sequence homology to endogenous transcripts from sequenced eukaryotes. Furthermore, they mimic properties of endogenous transcripts and cover a wide range of concentrations. To be able to use RNA spike-ins for normalization to global changes in gene expression, they have to be added to each sample in proportion to the number of cells (Lovén et al. 2012). Normalization factors can then be estimated by regression analysis of observed abundances in the RNA-seq data and known ERCCs input amounts. These normalization factors can then be plugged into differential gene expression analysis with DESeq2 as so-called scaling factors or edgeR as norm factors. Although ERCC spike-ins appear as a relatively simple and straightforward solution to the normalization problem for dual RNA-seq, there is still discussion how well they actually perform for normalization. It has been criticized that use of spike-ins requires high precision in adding them in the required concentrations (Robinson and Oshlack 2010), which is difficult to achieve, and that they do not actually mimic endogenous transcripts that well (Grün and van Oudenaarden 2015). Indeed, results by Risso et al. indicate that spike-ins are not reliable enough for use in standard global-scaling or regression-based normalization (Risso et al. 2014).

Furthermore, there is evidence that the mRNA enrichment method in RNA-seq library preparation, i.e., poly(A) selection or rRNA depletion, impacts spike-in read counts (Qing et al. 2013). Most importantly, while the behavior of spike-ins is at least fairly well understood for standard total RNA-seq, the combination with methods for purifying specific types of RNA, e.g., 4sU-labeled newly transcribed RNA or subcellular RNA fractions, has not yet been explored. As spike-ins can only be added subsequent to purification, the expected variability in purification yield between samples would make it extremely difficult if not impossible to add the correct quantities of spike-in RNAs. Thus, spike-in-based normalization is likely not reliable for these applications.

For virus-host transcriptomics, however—as for all dynamically changing systems—sequencing of newly transcribed (4sU-)RNA or nuclear or chromatin-associated RNA is more appropriate for DGE profiling than total RNA-seq since total RNA largely represents RNA transcribed before infection (Friedel and Dölken 2009, Dölken et al. 2008). Thus, even strong relative changes in transcription for a gene may not have a significant effect on total RNA levels during the duration of infection. Moreover, the effect on total RNA levels depends strongly on the basal turnover rates of corresponding genes, i.e., their transcription and RNA degradation rates in uninfected cells (Friedel and Dölken 2009). As a consequence, the identification of differentially expressed genes is biased toward genes with high basal RNA turnover. Furthermore, primary and secondary gene regulatory effects can be poorly distinguished as downstream differential expression of a gene with high RNA turnover may be detected before differential expression of a gene that is upstream in the regulatory cascade but has low RNA turnover. Even worse, in the presence of global effects such as the loss of host transcriptional activity in HSV-1 infection or increased mRNA degradation mediated by a virus protein, basal RNA turnover rates of genes influence how strongly their RNA levels are altered by these global effects (Friedel et al. 2021). In this case, differences in total RNA changes between genes resulting from differences in their RNA turnover rates can be mistaken for differences in susceptibility to the globally acting mechanisms. All of these problems can be mostly avoided by focusing on RNA transcribed only in short intervals of infection. While 4sU-seq or TT-seq is one possible way to achieve this, nuclear and in particular chromatin-associated RNA also represent RNA transcribed only very recently. Notably, we recently found chromatin-associated RNA to be particularly helpful for studying transcriptional regulation in HSV-1 infection without the confounding effects of both increased mRNA degradation and loss of host transcriptional activity (Friedel et al. 2021).

Although spike-in-based normalization is not an option for these methods, other ways could potentially be pursued to estimate normalization factors correcting for global gene expression changes. We previously attempted to use the fraction of rRNA reads in 4sU-seq samples for this purpose. Since both rRNA depletion and poly(A) selection are not applicable to 4sU-seq for the reasons noted above, a significant fraction of 4sU-seq reads originate from rRNA (up to 50%). Unfortunately, however, we found that rRNA content varied significantly between experiments and replicates, indicating that it was not a reliable quantity for normalization.

An alternative to correcting for global changes in gene expression by trying to find appropriate normalization factors is to focus on relative changes compared to all other genes rather than absolute changes. This can be easily achieved by using DGE analysis methods with their own built-in normalization. In this case, estimated fold-changes >1 do not necessarily represent an increase in RNA levels or transcription but—in presence of increased mRNA degradation or a general loss of transcription—simply less decrease than for other genes. Similarly, fold-changes <1 would indicate a higher reduction in RNA levels or transcription than for remaining genes. In virus-host transcriptomics, such relative changes may actually be more interesting as they are indicative of gene-specific regulation while absolute changes are dominated by global effects. For instance, in a recent comparison of wild-type (WT) HSV-1 infection with infection by a knock-out mutant of the HSV-1 virion host shutoff (*vhs*) protein, we found that *relative* fold-changes in chromatin-associated RNA were extremely well correlated at the same time-point of the infection (Friedel et al. 2021). Thus, although the mutant infection progressed more slowly, with a lower global reduction in host transcriptional activity at the same time-point than in WT infection, relative transcriptional changes were comparable for most genes. It should be noted, however, that the difference between absolute and relative changes has to be carefully explained in articles reporting on corresponding results, as it can otherwise lead to misunderstandings among the virologist audience.

9.5 BE CAREFUL OF YOUR INTERPRETATION

Apart from the challenges highlighted above, bioinformatics analysis of virus-host transcriptomic data is relatively straightforward. However, the interpretation of results is not always straightforward and may be misled by assumptions. This is of course not only the case in virus-host transcriptomics, but in any study of biological systems. One major cause of misinterpretation is the assumption that significant changes in gene read counts identified by DESeq2, edgeR, or similar methods indicate differential expression of the whole gene. However, what they really indicate is that there is a significant change in gene read counts, which may have other causes than actual differential expression. For instance, alternative splicing may lead to more frequent use of alternative transcripts that may be longer or shorter than the major isoform in the control condition, i.e., contain more or fewer exonic regions. This will either increase or reduce reads for the corresponding gene.

To identify alternative splicing, a number of software tools are also available with different approaches and objectives. DEXSeq aims at detecting differential exon usage by evaluating read counts for individual exons (Anders et al. 2012). In contrast, rMATS identifies alternative splicing events such as exon skipping/inclusion events using reads exclusive to corresponding isoforms (Shen et al. 2014). Both approaches have strengths and weaknesses. While rMATS is capable of detecting the type of alternative splicing, it requires a complete catalog of potential isoforms and will miss novel alternative splicing events. In contrast, DEXSeq can also pick up novel events that affect usage of (parts of) exons, but may be less sensitive to relatively small changes as it does not distinguish between exclusive and nonexclusive reads.

Both approaches are helpful for further interpreting results from DGE analysis and distinguishing differential expression of a gene from regulated alternative splicing. Nevertheless, significant changes in exon usage or splicing events also do not necessarily indicate "normal" alternative splicing but can be due to aberrant transcription changes not normally observed. An example for such aberrant transcription changes occurring in virus infections is the disruption of transcription termination observed in HSV-1 and influenza infection and some cellular stresses (Rutkowski et al. 2015, Vilborg et al. 2015, Bauer et al. 2018). Here, we were the first to identify disruption of transcription termination in any condition in our 4sU-seq time-course analysis of HSV-1 infection, but it was subsequently also identified in the other conditions. In the following, I will provide more details on how we first identified this phenomenon and the implications for bioinformatics analysis.

The original aim of our HSV-1 4sU-seq time-course experiment was two-fold: (i) characterize the host gene regulatory response to lytic HSV-1 infection and (ii) investigate splicing changes during infection as previous reports suggested that HSV-1 infection inhibits splicing (Hardy and Sandri-Goldin 1994, Sciabica et al. 2003). The initial DGE analysis indicated that most protein-coding genes were downregulated and only few (5.8%) were upregulated, which was consistent with our expectations. Surprisingly, however, when analyzing actively translated RNAs using ribosome profiling (Ingolia et al. 2009) we found that upregulation at transcriptional level barely lead to any translational upregulation. Only 0.34% of translated genes showed significant increased translational activity and the majority (77%) of transcriptionally upregulated genes were not translated at all. Even more remarkably, a surprisingly large fraction (44.5%) of annotated long intergenic non-coding RNAs (lincRNAs) appeared to be upregulated, most of which were only poorly characterized so far. This coincided with a general increase in reads mapping to intergenic and intronic regions compared to exonic regions. While the increase in intronic reads appeared to confirm the general inhibition of splicing in HSV-1 infection, the other results raised our suspicion that some other yet unknown mechanism was at play.

When looking at mapped reads in a genome browser for genes that were transcriptionally upregulated but not translated, we saw that there was indeed increased transcriptional activity in those genes. However, this transcriptional activity did not appear to originate at the gene promoter but at an upstream gene, continuing for tens-of-thousands of nucleotides beyond its poly(A) site into the intergenic space and into downstream genes (see Figure 9.3a). This read-through transcription, as we termed it, affected a substantial fraction of host genes. At the end of our time-course (7–8 hours p.i.), 64% of ~10,000 expressed host genes showed read-through transcription of more than 15% of their expression and 26% showed more than 75%. While the so-called "read-in" transcription from an upstream gene affected a smaller fraction of genes (32.6% of genes with read-in transcription >15%), this read-in transcription explained the increase in intronic reads as introns of read-in transcripts were often (but not always) not properly spliced. Genes without read-in transcription showed no splicing defects, arguing against a general inhibition of cotranscriptional splicing by HSV-1. Strikingly, we even observed intergenic splicing between exons of genes in close proximity that were connected by read-though transcription. These intergenic splicing events often superseded "normal" splicing events that the corresponding splice sites were involved in uninfected cells. This not only provided evidence that read-through transcription

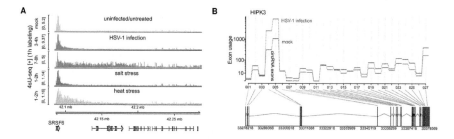

FIGURE 9.3 (a) Read-through transcription in HSV-1 infection and salt and heat stress for the example gene SRSF6. Shown are mapped sequencing reads (average of two replicates after normalizing to sequencing depth) from 4sU-seq data (with 1 hour labeling) for indicated time-points in HSV-1 infection and stresses. Genes are shown in the bottom track. Read-through transcription for the SRSF6 gene leads to more or less extensive intergenic transcription—depending on the condition—that can extend into downstream genes and lead to seeming "induction" of these genes. (b) Escape of circRNAs from virus-induced mRNA degradation is identified as differential exon usage of corresponding exons. DEXSeq results comparing total RNA-seq data in mock and HSV-1 infection (8 hours p.i.) are shown for an example gene (HIPK3). This identifies increased usage in HSV-1 infection compared to mock for one exon (split into two parts due to its use in alternative transcripts) that is not only part of linear HIPK3 transcripts but also part of a circRNA.

represented continuous transcription from upstream genes rather than repeated transcription initiation, but also has interesting implications regarding the kinetics of cotranscriptional splicing.

Read-in transcription also explained the seeming induction of so many lincRNAs as well as protein-coding genes without detectable translation. Since most of these were either not or only poorly expressed, read-in transcription easily exceeded their endogenous levels. Analysis of subcellular RNA fractions showed that read-through transcripts are not exported from the nucleus, providing more evidence that read-in transcription does not lead to any functional upregulation of these genes. Moreover, this suggests that read-through transcription effectively leads to functional downregulation of genes upstream of affected poly(A) sites as newly transcribed RNAs are retained in the nucleus and cannot be translated to protein.

These observations have two major implications for bioinformatics analysis of virus-host transcriptomic data. First, as read-through was only recently discovered in such well-studied processes as salt and heat stress and has been confirmed for two viruses already, i.e., HSV-1 and influenza, it is likely to occur in other (stress) conditions and infections. Thus, it should always be evaluated in any dual RNA-seq data analysis. Otherwise, there is a danger of mistaking read-through transcription for "normal" alternative splicing or polyadenylation. A case in point for this problem is a recent study that reported significant changes in alternative splicing, in particular increased intron retention, and alternative polyadenylation in HSV-1 infection (Hu et al. 2016). Comparison of their results to ours, however, indicated that many (but not all) of the identified alternative splicing and polyadenylation events follow from read-in and read-through transcription, respectively. It should be noted that read-through is distinct from "normal" alternative polyadenylation as reads tend to decrease gradually with increasing distance from the poly(A) site (see Figure 9.3a). In contrast, proper alternative polyadenylation is reflected in comparably sharp read drop-offs at the 3′ end of transcripts.

In our original article, we quantified the extent of read-through and read-in transcription for individual genes from 4sU-seq data by calculating the ratio of expression in a 5 kb window down- or upstream of the gene, respectively, divided by expression of the gene itself. To quantify expression for the 5 kb windows or the gene itself any metric for quantifying expression from gene read counts can be used. Commonly used ones are RPKM (Reads Per Kilobase Million), FPKM (Fragments Per Kilobase Million, for paired-end sequencing data), or TPM (Transcripts Per Kilobase Million). The only difference between R/FPKM on the one hand and TPM on the other hand is the scaling factors that normalize for sequencing depth. However, since this is the same for the gene and down- or upstream windows, it will be cancelled out in the ratio. Thus, use of either R/FPKM or TPM will lead to the same read-through values. This has the additional advantage that no normalization factors have to be determined to account for changes in global RNA levels or host transcriptional activity. One problem, however, that remains with this definition of read-through is that background noise down- or upstream of genes may lead to artificially high read-through values for lowly expressed genes even in uninfected cells. In our original time-course analysis, we circumvented this problem by requiring also a consistent increase in read-through or read-in values throughout infection; however, this is not practical if only one or a few time-points of infection are probed. We thus developed an alternative approach in a more recent study by subtracting values in uninfected cells from values in infected cells (Hennig et al. 2018), with negative values set to zero.

A different approach was developed by Vilborg et al. in their study of read-through transcription upon cellular stresses, which they originally termed Downstream of Gene containing transcripts (DoGs) (Vilborg et al. 2015). To identify candidate DoGs, they required >95% coverage by reads in the 5 or 10 kb downstream of genes, with coverage defined as the percentage of nucleotide positions covered by at least one read. For candidate DoGs, they then determined the RPKM in the last 1 kb of the gene and for every 1 kb window downstream of genes until the RPKM of a downstream window dropped below 1% of the RPKM of the last 1 kb of the gene. This defined the endpoints of candidate DoGs in untreated and stress-treated conditions. An increase in length from untreated to treated conditions was then used to identify the actual DoGs. While this approach has the advantage that it actually predicts an endpoint for read-through transcription, it is highly sensitive both to sequencing depth and expression of genes. Thus, in a more recent study, they performed subsampling of reads for all samples down to the lowest number of aligned reads observed in any sample before predicting DoGs (Vilborg et al. 2017). In contrast, our method is mostly independent of both sequencing depth and gene expression, although variance of estimates obviously increases with lower sequencing depth and gene expression. Furthermore, both our studies and the studies by Vilborg et al. indicated that there is no clearly defined endpoint to read-through transcription as read coverage tends to drop continuously downstream of genes. In addition, both the extent and length of read-through commonly increases with the duration of infection or stress. Thus, computationally defined endpoints of read-through will largely be arbitrary. Interestingly, however, we found that our measure of read-through transcription was linearly correlated to the variance in host gene expression changes in both HSV-1 infection and salt and heat stress (Wang et al. 2020, Hennig et al. 2018), suggesting that the extent of perturbation of the cell may determine the extent of read-through in these conditions. This also allowed determining an HSV-1 protein that is at least partly responsible for inducing read-through transcription (Wang et al. 2020). While a knockout mutant of this protein still induced read-through, the extent was much lower than expected from the variance of observed host gene expression changes.

The second implication of read-through and in particular read-in transcription is that great care has to be taken when selecting the set of genes for DEG analysis in affected conditions. Since read-in transcription does not lead to functional upregulation of genes but can easily be mistaken for this, genes with significant amounts of read-in transcription should first be excluded from any DGE analysis. Otherwise, functional interpretation of results may be biased by the seeming induction of some genes by read-through. Since lowly expressed genes appear particularly strongly induced by read-in transcription, they would likely be at the top of the list of upregulated genes. One approach to avoid this problem could be to perform the DGE analysis only on our previously defined set of ~4000 genes without read-in transcription in either HSV-1 infection or salt and heat stress (Hennig et al. 2018). However, that may be unnecessarily stringent as read-through transcription and consequently read-in transcription may be lower in the specific condition analyzed. Furthermore, while there was a significant overlap between genes with read-through in HSV-1 infection and salt and heat stress, respectively, there were also considerable differences between all three conditions. Thus, it is recommendable to redefine genes not affected by read-in transcription for every experiment in which read-through transcription is observed.

Read-through transcription is not the only phenomenon that can bias DGE and alternative splicing analyses. For instance, we previously observed the opposite effect, i.e., premature transcription termination, upon CDK12 inhibition, which we first observed as downregulation of genes in RNA-seq data of transcript 3′ ends (Chirackal Manavalan et al. 2019). Only more in-depth analyses and additional whole-transcript nuclear RNA-seq data revealed this to be due to shortening of transcripts rather than overall downregulation of expression of these genes. Although this experiment did not involve a virus infection, similar effects could occur in some virus infections and should at least be considered.

There are also other examples not involving transcription termination where the wrong interpretation of results can lead to wrong conclusions. This is highlighted by the HSV-1 *vhs* protein, the HSV-1 protein that targets host and viral mRNA for degradation. Although *vhs* is an endonuclease, it is targeted to mRNAs in a translation-initiation-dependent manner (Feng et al. 2001) and does not cleave circular RNAs (circRNAs) unless they contain an internal ribosome entry site (Shiflett and Read 2013). CircRNAs form closed RNA-loops, in which 5′ and 3′ ends are covalently joined, and are naturally generated during the splicing process either by "back-splicing" of exons out of their linear order or stabilization of intron lariats (Vicens and Westhof 2014, Lasda and Parker 2014). Previously considered to occur only rarely (Nigro et al. 1991), recent large-scale RNA-seq studies revealed the existence of thousands of circRNAs in eukaryotic cells (Salzman et al. 2012, Jeck et al. 2013, Memczak et al. 2013, Wang et al. 2014). Since naturally occurring circRNAs should be resistant to *vhs*-mediated degradation, they are expected to accumulate in HSV-1 infection relative to linear transcripts. Indeed, Shi et al. recently reported upregulation of 188 circRNAs at 48 hours p.i. HSV-1 infection (Shi et al. 2018), but interpreted this as dysregulation of circRNAs in HSV-1 infection rather than as evidence of their escape from *vhs*-mediated RNA degradation. Consistent with this, alternative splicing analysis with DEXseq on total RNA-seq data of HSV-1 infection obtained in the same experiment as our 4sU-seq time-course finds increased usage for exons that are part of a circRNA compared to exons of the same gene that are only part of linear transcripts. This is shown for an example gene in Figure 9.3b. Without careful analysis, this might also be mistaken as evidence of alternative splicing. Interestingly, almost no circRNAs are found in 4sU-seq data, which might be reflective of the very low synthesis rate of circRNAs or indicate problems in biochemical separation of 4sU-labeled circRNAs.

A different example of how assumptions can bias the interpretation of results in virus-host transcriptomics can be found in a recent total RNA-seq study of HSV-1 infection by Pheasant et al. (2018). They noted a high variability in the extent of downregulation in absolute mRNA levels between host protein-coding genes and hypothesized that this might be due to differences in susceptibility to *vhs*-mediated degradation. They confirmed high *vhs*-dependent downregulation for a few genes by PCR and showed that this was not due to lower transcription rates or basal RNA turnover of these genes compared to a few genes that showed less reduction in RNA levels. From this, they concluded that the differences in *vhs*-dependent downregulation between genes indicated differences in susceptibility of their mRNAs to *vhs*-mediated degradation. This assumed that *vhs* only affects stability of transcripts and has no other direct or indirect effects on mRNA levels. More recently, we found, however, that a subset of genes is also transcriptionally downregulated in a *vhs*-mediated manner (Friedel et al. 2021), including the genes with high reduction in total mRNA levels tested by Pheasant et al. (2018). Thus, the combined effects of *vhs*-mediated mRNA degradation and transcriptional downregulation explain the strong reduction of their mRNA levels without requiring any differences in susceptibility to degradation. For other genes, the variability can at least in parts be explained by differences in basal RNA turnover.

9.6 CONCLUSION

In many regards, bioinformatics analysis of virus-host transcriptome data obtained with dual RNA-seq does not differ from the analysis applied to other RNA-seq experiments. In particular, the problem of normalization occurs in many conditions with global changes in gene expression. Unique challenges arise mostly from the poor gene annotation of most viral genomes, inconsistencies between different approaches for annotating genomes, and the lack of a general database for depositing updated annotations in standard file formats. When interpreting the results from standard bioinformatics analysis on dual RNA-seq data, however, it is important to always consider that there may be other unsuspected mechanisms that could lead to the observed results instead of the most obvious ones or the ones most consistent with published literature. A lot of established "knowledge" in virology is based on small-scale studies on a few genes that were often performed in vitro. Moreover, it is commonly the interpretation of results that is carried on via citation chains rather than the actual experimental evidence. For instance, most articles on *vhs* note in the introduction that it is inactivated later in infection. However, when following the chain of citations to their origin, one only finds an article that shows dampening of *vhs* activity not complete inactivation (Lam et al. 1996). Thus, it can be helpful for a bioinformatician performing virus-host transcriptome data analysis to first check the evidence that underlies some "long-known facts," in particularly if they are contradictory to what the dual RNA-seq data suggest.

REFERENCES

Abrisch, R. G., T. M. Eidem, P. Yakovchuk, J. F. Kugel, and J. A. Goodrich. 2016. "Infection by herpes simplex virus 1 causes near-complete loss of RNA polymerase II occupancy on the host cell genome." *J Virol* 90 (5):2503–13.

Anders, S., A. Reyes, and W. Huber. 2012. "Detecting differential usage of exons from RNA-seq data." *Genome Res* 22 (10):2008–17.

Arias, C., B. Weisburd, N. Stern-Ginossar, et al. 2014. "KSHV 2.0: A comprehensive annotation of the Kaposi's sarcoma-associated herpesvirus genome using next-generation sequencing reveals novel genomic and functional features." *PLoS Pathog* 10 (1):e1003847.

Bauer, D. L. V., M. Tellier, M. Martínez-Alonso, et al. 2018. "Influenza virus mounts a two-pronged attack on host RNA polymerase II transcription." *Cell Rep* 23 (7):2119–29.e3.

Bencun, M., O. Klinke, A. Hotz-Wagenblatt, et al. 2018. "Translational profiling of B cells infected with the Epstein-Barr virus reveals 5′ leader ribosome recruitment through upstream open reading frames." *Nucleic Acids Res* 46 (6):2802–19.

Benson, D. A., M. Cavanaugh, K. Clark, et al. 2013. "GenBank." *Nucleic Acids Res* 41 (Database issue):D36–42.

Birkenheuer, C. H., C. G. Danko, and J. D. Baines. 2018. "Herpes simplex virus 1 dramatically alters loading and positioning of RNA polymerase II on host genes early in infection." *J Virol* 92 (8):e02184–17.

Blanco-Melo, D., B. E. Nilsson-Payant, W.-C. Liu, et al. 2020. "Imbalanced host response to SARS-CoV-2 drives development of COVID-19." *Cell* 181 (5):1036–45.e9.

Blinov, V. M., V. V. Zverev, G. S. Krasnov, F. P. Filatov, and A. V. Shargunov. 2017. "Viral component of the human genome." *Mol Biol* 51 (2):205–15.

Bonfert, T., G. Csaba, R. Zimmer, and C. C. Friedel. 2013. "Mining RNA-seq data for infections and contaminations." *PLoS One* 8 (9):e73071.

Bonfert, T., E. Kirner, G. Csaba, R. Zimmer, and C. C. Friedel. 2015. "ContextMap 2: Fast and accurate context-based RNA-seq mapping." *BMC Bioinf* 16:122.

Bray, N. L., H. Pimentel, P. Melsted, and L. Pachter. 2016. "Near-optimal probabilistic RNA-seq quantification." *Nat Biotechnol* 34 (5):525–7.

Cao, Y., K. Zhang, L. Liu, et al. 2019. "Global transcriptome analysis of H5N1 influenza virus-infected human cells." *Hereditas* 156:10.

Chee, M. S., A. T. Bankier, S. Beck, et al. 1990. "Analysis of the protein-coding content of the sequence of human cytomegalovirus strain AD169." *Curr Top Microbiol Immunol* 154:125–69.

Chirackal Manavalan, A. P., K. Pilarova, M. Kluge, et al. 2019. "CDK12 controls G1/S progression by regulating RNAPII processivity at core DNA replication genes." *EMBO Rep* 20 (9):e47592.

Core, L. J., J. J. Waterfall, and J. T. Lis. 2008. "Nascent RNA sequencing reveals widespread pausing and divergent initiation at human promoters." *Science (New York, N.Y.)* 322 (5909):1845–8.

Davison, A. J., A. Dolan, P. Akter, et al. 2003. "The human cytomegalovirus genome revisited: Comparison with the chimpanzee cytomegalovirus genome." *J Gen Virol* 84 (Pt 1):17–28.

Dobin, A., C. A. Davis, F. Schlesinger, et al. 2013. "STAR: Ultrafast universal RNA-seq aligner." *Bioinformatics* 29 (1):15–21.

Dölken, L., Z. Ruzsics, B. Radle, et al. 2008. "High-resolution gene expression profiling for simultaneous kinetic parameter analysis of RNA synthesis and decay." *RNA* 14 (9):1959–72.

Feng, P., D. N. Everly, and G. S. Read. 2001. "mRNA decay during herpesvirus infections: Interaction between a putative viral nuclease and a cellular translation factor." *J Virol* 75 (21):10272–80.

Friedel, C. C., and L. Dölken. 2009. "Metabolic tagging and purification of nascent RNA: Implications for transcriptomics." *Mol Biosyst* 5 (11):1271–8.

Friedel, C. C., A. W. Whisnant, L. Djakovic, et al. 2021. "Dissecting Herpes Simplex Virus 1-Induced Host Shutoff at the RNA Level." *J Virol* 95 (3):e01399–20.

Glaunsinger, B., and D. Ganem. 2004. "Lytic KSHV infection inhibits host gene expression by accelerating global mRNA turnover." *Mol Cell* 13 (5):713–23.

Grün, D., and A. van Oudenaarden. 2015. "Design and analysis of single-cell sequencing experiments." *Cell* 163 (4):799–810.

Hardy, W. R., and R. M. Sandri-Goldin. 1994. "Herpes simplex virus inhibits host cell splicing, and regulatory protein ICP27 is required for this effect." *J Virol* 68 (12):7790–9.

Hennig, T., M. Michalski, A. J. Rutkowski, et al. 2018. "HSV-1-induced disruption of transcription termination resembles a cellular stress response but selectively increases chromatin accessibility downstream of genes." *PLoS Pathog* 14 (3):e1006954.

Hu, B., X. Li, Y. Huo, et al. 2016. "Cellular responses to HSV-1 infection are linked to specific types of alterations in the host transcriptome." *Sci Rep* 6 (1):28075.

Ingolia, N. T., S. Ghaemmaghami, J. R. S. Newman, and J. S. Weissman. 2009. "Genome-wide analysis in vivo of translation with nucleotide resolution using ribosome profiling." *Science* 324 (5924):218–23.

Jagger, B. W., H. M. Wise, J. C. Kash, et al. 2012. "An overlapping protein-coding region in influenza A virus segment 3 modulates the host response." *Science* 337 (6091):199–204.

Jeck, W. R., J. A. Sorrentino, K. Wang, et al. 2013. "Circular RNAs are abundant, conserved, and associated with ALU repeats." *RNA* 19 (2):141–57.

Jiang, L., F. Schlesinger, C. A. Davis, et al. 2011. "Synthetic spike-in standards for RNA-seq experiments." *Genome Res* 21 (9):1543–51.

Kamitani, W., C. Huang, K. Narayanan, K. G. Lokugamage, and S. Makino. 2009. "A two-pronged strategy to suppress host protein synthesis by SARS coronavirus Nsp1 protein." *Nat Struct Mol Biol* 16 (11):1134–40.

Kim, D., B. Langmead, and S. L. Salzberg. 2015. "HISAT: A fast spliced aligner with low memory requirements." *Nat Methods* 12 (4):357–60.

Kwong, A. D., and N. Frenkel. 1987. "Herpes simplex virus-infected cells contain a function(s) that destabilizes both host and viral mRNAs." *Proc Natl Acad Sci U S A* 84 (7):1926–30.

Lam, Q., C. A. Smibert, K. E. Koop, et al. 1996. "Herpes simplex virus VP16 rescues viral mRNA from destruction by the virion host shutoff function." *EMBO J* 15 (10):2575–81.

Lasda, E., and R. Parker. 2014. "Circular RNAs: Diversity of form and function." *RNA* 20 (12):1829–42.

Love, M. I., W. Huber, and S. Anders. 2014. "Moderated estimation of fold change and dispersion for RNA-seq data with DESeq2." *Genome Biol* 15 (12):550.

Lovén, J., David A. Orlando, Alla A. Sigova, et al. 2012. "Revisiting global gene expression analysis." *Cell* 151 (3):476–82.

Maeda, E., M. Akahane, S. Kiryu, et al. 2009. "Spectrum of Epstein-Barr virus-related diseases: A pictorial review." *Jpn J Radiol* 27 (1):4–19.

Mahat, D. B., H. Kwak, G. T. Booth, et al. 2016. "Base-pair-resolution genome-wide mapping of active RNA polymerases using precision nuclear run-on (PRO-seq)." *Nat Protoc* 11 (8):1455–76.

Marcinowski, L., M. Lidschreiber, L. Windhager, et al. 2012. "Real-time transcriptional profiling of cellular and viral gene expression during lytic cytomegalovirus infection." *PLoS Pathog* 8 (9):e1002908.

Memczak, S., M. Jens, A. Elefsinioti, et al. 2013. "Circular RNAs are a large class of animal RNAs with regulatory potency." *Nature* 495 (7441):333–8.

Murphy, E., D. Yu, J. Grimwood, et al. 2003b. "Coding potential of laboratory and clinical strains of human cytomegalovirus." *Proc Natl Acad Sci U S A* 100 (25):14976–81.

Murphy, E., I. Rigoutsos, T. Shibuya, and T. E. Shenk. 2003a. "Reevaluation of human cytomegalovirus coding potential." *Proc Natl Acad Sci U S A* 100 (23):13585–90.

Nigro, J. M., K. R. Cho, E. R. Fearon, et al. 1991. "Scrambled exons." *Cell* 64 (3):607–13.

Patro, R., G. Duggal, M. I. Love, R. A. Irizarry, and C. Kingsford. 2017. "Salmon provides fast and bias-aware quantification of transcript expression." *Nat Methods* 14 (4):417–9.

Pellett, P. E., and B. Roizman. 2013. "Herpesviridae." In *Fields Virology*, edited by D. M. Knipe and P. M. Howley, pp. 1802–21.Philadelphia, PA: Lippincott Williams & Wilkins.

Pheasant, K., C. S. Möller-Levet, J. Jones, et al. 2018. "Nuclear-cytoplasmic compartmentalization of the herpes simplex virus 1 infected cell transcriptome is co-ordinated by the viral endoribonuclease vhs and cofactors to facilitate the translation of late proteins." *PLoS Pathog* 14 (11):e1007331.

Qing, T., Y. Yu, T. Du, and L. Shi. 2013. "mRNA enrichment protocols determine the quantification characteristics of external RNA spike-in controls in RNA-Seq studies." *Sci China Life Sci.* 56 (2):134–42.

Risso, D., J. Ngai, T. P. Speed, and S. Dudoit. 2014. "Normalization of RNA-seq data using factor analysis of control genes or samples." *Nat Biotechnol* 32 (9):896–902.

Robinson, M. D., and A. Oshlack. 2010. "A scaling normalization method for differential expression analysis of RNA-seq data." *Genome Biol* 11 (3):R25.

Robinson, M. D., D. J. McCarthy, and G. K. Smyth. 2009. "edgeR: A bioconductor package for differential expression analysis of digital gene expression data." *Bioinformatics* 26 (1):139–40.

Roizman, B., D. M. Knipe, and R. J. Whitley. 2007. "Herpes simplex viruses." In *Fields Virology*, 5th edn, edited by P. M. Howley and D. M. Knipe, pp. 2501–601. Philadelphia, PA: Lippincott Williams & Wilkins.

Rowe, M., B. Glaunsinger, D. van Leeuwen, et al. 2007. "Host shutoff during productive Epstein-Barr virus infection is mediated by BGLF5 and may contribute to immune evasion." *Proc Natl Acad Sci U S A* 104 (9):3366–71.

Rutkowski, A. J., F. Erhard, A. L'Hernault, et al. 2015. "Widespread disruption of host transcription termination in HSV-1 infection." *Nat Commun* 6:7126.

Salzman, J., C. Gawad, P. L. Wang, N. Lacayo, and P. O. Brown. 2012. "Circular RNAs are the predominant transcript isoform from hundreds of human genes in diverse cell types." *PLoS One* 7 (2):e30733.

Schwalb, B., M. Michel, B. Zacher, et al. 2016. "TT-seq maps the human transient transcriptome." *Science* 352 (6290):1225–28.

Sciabica, K. S., Q. J. Dai, and R. M. Sandri-Goldin. 2003. "ICP27 interacts with SRPK1 to mediate HSV splicing inhibition by altering SR protein phosphorylation." *The EMBO J* 22 (7):1608–19.

Shen, S., J. W. Park, Z.-X. Lu, et al. 2014. "rMATS: Robust and flexible detection of differential alternative splicing from replicate RNA-Seq data." *Proc Natl Acad Sci* 111 (51):E5593–601.

Shi, J., N. Hu, L. Mo, et al. 2018. "Deep RNA sequencing reveals a repertoire of human fibroblast circular RNAs associated with cellular responses to herpes simplex virus 1 infection." *Cell Physiol Biochem* 47 (5):2031–45.

Shiflett, L. A., and G. S. Read. 2013. "mRNA decay during Herpes Simplex Virus (HSV) infections: Mutations that affect translation of an mRNA influence the sites at which it is cleaved by the HSV Virion Host Shutoff (Vhs) protein." *J Virol* 87 (1):94–109.

Soon, W. W., M. Hariharan, and M. P. Snyder. 2013. "High-throughput sequencing for biology and medicine." *Mol Syst Biol* 9 (1):640.

Stern-Ginossar, N., B. Weisburd, A. Michalski, et al. 2012. "Decoding human cytomegalovirus." *Science* 338 (6110):1088–93.

Varnum, S. M., D. N. Streblow, M. E. Monroe, et al. 2004. "Identification of proteins in human cytomegalovirus (HCMV) particles: The HCMV proteome." *J Virol* 78 (20):10960–6.

Vicens, Q., and E. Westhof. 2014. "Biogenesis of circular RNAs." *Cell* 159 (1):13–14.

Vilborg, A., M. C. Passarelli, T. A. Yario, K. T. Tycowski, and J. A. Steitz. 2015. "Widespread inducible transcription downstream of human genes." *Mol Cell* 59 (3):449–61.

Vilborg, A., N. Sabath, Y. Wiesel, et al. 2017. "Comparative analysis reveals genomic features of stress-induced transcriptional readthrough." *Proc Natl Acad Sci* 114 (40):E8362–71.

Wang, P. L., Y. Bao, M. C. Yee, et al. 2014. "Circular RNA is expressed across the eukaryotic tree of life." *PLoS One* 9 (6):e90859.

Wang, Z., M. Gerstein, and M. Snyder. 2009. "RNA-Seq: A revolutionary tool for transcriptomics." *Nat Rev Genet* 10 (1):57–63.

Wang, X., T. Hennig, A. W. Whisnant, et al. 2020. "Herpes simplex virus blocks host transcription termination via the bimodal activities of ICP27." *Nat Commun* 11 (1):293.

Weidner-Glunde, M., E. Kruminis-Kaszkiel, and M. Savanagouder. 2020. "Herpesviral latency-common themes." *Pathogens* 9 (2):125.

Westermann, A. J., S. A. Gorski, and J. Vogel. 2012. "Dual RNA-seq of pathogen and host." *Nat Rev Microbiol* 10 (9):618–30.

Whisnant, A. W., C. S. Jurges, T. Hennig, et al. 2020. "Integrative functional genomics decodes herpes simplex virus 1." *Nat Commun* 11 (1):2038.

Windhager, L., T. Bonfert, K. Burger, et al. 2012. "Ultrashort and progressive 4sU-tagging reveals key characteristics of RNA processing at nucleotide resolution." *Genome Res* 22 (10):2031–42.

Wrensch, M., A. Weinberg, J. Wiencke, et al. 2001. "Prevalence of antibodies to four herpesviruses among adults with glioma and controls." *Am J Epidemiol* 154 (2):161–5.

Yates, A. D., P. Achuthan, W. Akanni, et al. 2019. "Ensembl 2020." *Nucleic Acids Res* 48 (D1):D682–8.

Yu, D., M. C. Silva, and T. Shenk. 2003. "Functional map of human cytomegalovirus AD169 defined by global mutational analysis." *Proc Natl Acad Sci U S A* 100 (21):12396–401.

CHAPTER 10

Sequence Classification with Machine Learning at the Example of Viral Host Prediction

Florian Mock and Manja Marz

Friedrich Schiller University Jena

CONTENTS

10.1 MACHINE LEARNING APPLICATIONS IN VIROLOGY

Machine learning is a commonly used term to describe algorithms that can solve tasks based on data provided without being explicitly programmed for this task. Rather, only concepts of data analysis and transformation are

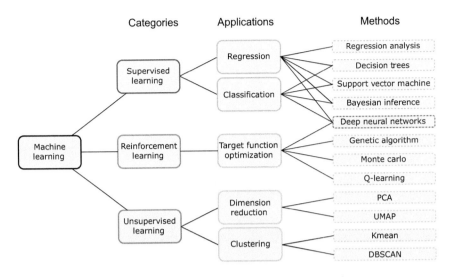

FIGURE 10.1 Overview of different machine learning categories, applications, and methods.

programmed. Machine learning has the advantage that less initial knowledge about the problem is needed. Instead, the program learns patterns and regularities in the data that help to solve the problem.

The field of machine learning is broadly divided into three different categories, namely reinforcement, supervised, and unsupervised machine learning (see Figure 10.1).

For *reinforcement machine learning*, an algorithm is trained by rewarding or punishing previous decisions. This method is used for dynamic environments with which the algorithm interacts to reach a predefined goal. Reinforcement learning is widely used in recommender systems but also finds application in systems biology. Nevertheless, reinforcement learning in bioinformatics is rather a niche phenomenon. In the field of virology, it is striking that reinforcement machine learning is used primarily for modeling infection events. For example, Monte Carlo, which is basically a repeated random sampling to obtain numerical results, was already used in 1988 to model the spread of human immunodeficiency virus (HIV) (Barrett, 1988). The evolution of HIV virus was modeled using a genetic algorithm (Bocharov et al., 2005). A genetic algorithm uses a population of possible solutions, selects the best performing one, and produces offspring. The offspring is a group of replicates of the best performing solution with small random changes. This method is repeated multiple times. Q-learning was used for modeling the ideal drug delivery rate to reduce the infected cells and free virus particles (virions) of HIV (Gholizade-Narm and Noori, 2018). In general, Q-learning is an action learning technique that learns the quality of actions depending on the circumstances and can, therefore, be used to suggest actions. In reinforcement learning, the use of deep neural networks is also known as deep reinforcement. Deep reinforcement, like Q-learning, learns to evaluate actions depending on the circumstances and can even develop sophisticated long-term strategies. This method was used to model epidemiological prevention strategies for COVID-19 (Libin et al., 2020).

Unsupervised machine learning is mainly used to identify patterns and regularities in the data. These patterns are often used for data clustering or as a data reduction technique. Unsupervised machine learning algorithms do rely solely on the statistical properties of the given input samples. These algorithms are distinguished into two main groups, namely dimension reduction algorithms and clustering algorithms.

Dimension reduction algorithms are mainly used to simplify the representation of complex data with a minimal loss of information. Among others, principal component analysis (PCA) and Uniform Manifold Approximation and Projection (UMAP) (Wold et al., 1987; McInnes et al., 2018) are commonly used for this purpose. PCA maximizes the represented data's variance by combining the dimensions (see Figure 10.2b). This preserves the pairwise distance structure amongst all data samples. Therefore, the similarity of samples can be inferred by the distance in the PCA plot. This method is commonly used to visualize complex data (Gerst and Hölzer, 2018); for example, Raman spectroscopy results for influenza A virus-infected cells (Lim et al., 2019). In contrast to PCA, UMAP favors the preservation of local distances over global distances; this helps to represent even more complex relationships between samples. UMAP was, for example, used for analyzing the mutations of severe acute respiratory syndrome coronavirus 2 (SARS-CoV-2) (Hozumi et al., 2020).

Clustering algorithms label samples into groups based on statistical similarity. Two common clustering algorithms are Kmeans and DBSCAN (Steinhaus, 1956; Ester et al., 1996). Kmeans is a classic approach that clusters the whole data into a user-defined number of k clusters. Kmeans helped to identify the significant changes across the four clinical phases of chronic hepatitis B virus (HBV) infection (Schoeman et al., 2016). In contrast to Kmeans, DBSCAN automatically identifies the number of clusters and can even identify noise. Therefore, it is often more applicable to real-world data. DBSCAN was used to study the origin of coronavirus disease 2019 (COVID-19) (Nguyen et al., 2020).

Supervised machine learning is based on a training step on a set of training samples including the expected output for each sample. The algorithm learns to recognize patterns in the input data to transform/map it to the output data. This type of machine learning is called supervised, as the expected output is provided to method. A typical application is the classification of the input data into different categories (Kotsiantis et al., 2007), for example the classification of a genome fragment into categories describing its taxonomic origin, such as virus, bacteria, or eukaryota (Kretschmer et al. 2021). Regression analysis is a commonly used technique to identify data trends by fitting a curve or distribution to the data. This method was already used in the 1960s to enhance the estimation of the infection level of pox (Gart, 1964).

MACHINE LEARNING:
- *Feature*: A measurable property of a sample. For example, the length of a genome.
- *Training set*: A set of multiple samples, which is used for training the algorithm. For example, the training set could consist of several DNA sequences.
- *Validation set*: A smaller set which is used to decide when to stop the training. For this, we predict the validation set and check the quality of the prediction. The validation set consists of the same data type as the training set, for example, DNA sequences.
- *Test set*: After the completed training, we evaluate the quality of the trained algorithm on the test set.

Note that the training, validation, and test sets do not have any samples in common.

GRAPH THEORY (SEE FIGURE 10.2A):
- *Node*: The basic unit of a graph.
- *Edge*: An edge connects two nodes.
- *Path*: A path is a sequence of nodes and edges.
- *Tree*: A tree is a graph in which any two nodes are connected by exactly one path.
- *Rooted tree*: A rooted tree is a tree in which one node has been designated the root.
- *Leaf*: All nodes which are connected by only one edge.

Supervised and unsupervised machine learning are broadly used in bioinformatics. Both techniques free the user from defining usable thresholds and parameters and can represent more complex regularities found in biology. Furthermore, as these approaches are data-driven, they significantly reduce the self-fulfilling prophecy problematic.

In this chapter, we are particularly interested in classification methods. To go into more detail, we introduce a few terms.

Decision trees use a tree structure (see Figure 10.2a), for which each internal node is labeled with an input feature, and each leaf represents a class. During classification, a binary decision is made at each node, depending on the queried feature value. With this, the sample iterates downward in the tree until a leaf is reached. Decision trees have the advantage that they are easy to understand after training. Meaning, the decision process is clear, and it is simple to determine which features are important. Multiple different styles of decision trees exist, such as Random Forests or AdaBoost. Random Forests use multiple trees, each based on a subset of the input features, to predict the class. The class which was predicted by most trees is the prediction of the Random Forests. In contrast to Random Forests, AdaBoost does not use fully sized decision trees but stumps, which consist only of a root and two leaves. Furthermore, AdaBoost builds these stumps consecutively. Each stump is build to reduce the error of the previous stumps. Next, the new stump is weighted depending on how much the error is reduced. During prediction, all stumps individually predict

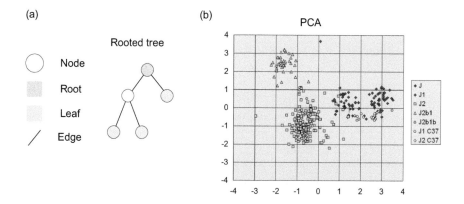

FIGURE 10.2 The tree graph (a) is a commonly used structure for different machine learning approaches, such as decision trees. The principal component analysis (b) is likely the most commonly used dimension reduction method. (Panel b is provided by Commons (2020).)

the sample, like the trees in Random Forests. Unlike for Random Forests, the class with the highest sum of weights over all stumps rather than the most common predicted class is the output of AdaBoost. Decision trees are commonly used in bioinformatics tasks such as diagnosing dengue virus-based microscopic images (Tantikitti et al., 2015) or identifying biomarkers for early diagnosis of HBV-related liver fibrosis (Lu et al., 2010).

In a *Support Vector Machine* (SVM), the samples are represented as points in space. Each dimension in space corresponds to a specific feature of the input. Thus, the feature expression of a sample defines the point in space. An SVM represents the points in space so that the broadest possible gap separates the training samples of the individual classes. To divide all classes, hyperplanes are placed through this gap. New samples will be mapped into the same space and are predicted according to the hyperplane sides of their placement. SVMs can also predict nonlinear problems; for this, the so-called kernel trick is used for which a higher number of dimensions than in the input data is introduced before defining the hyperplanes. The ability to predict nonlinear problems is a clear advantage compared to decision trees. SVMs are used in a wide range of applications in bioinformatics, for example, the identification of potentially important gene (Metzler and Kalinina, 2014) and protein functions (Sahoo et al., 2008; Modak et al., 2019) in virology. Deciding if an SVM or a decision tree approach is more feasible for the problem of interest is hard. Suggestions can be made depending on the problem and data of interest. Generally, SVMs tend to achieve higher areas under the curve (AUC) than decision trees (Huang et al., 2003). Problems with a high feature count favor Random Forests (Yang et al., 2010). Furthermore, decision trees are easier to interpret. It is advisable to test both methods for the problem of interest.

10.2 AN INTRODUCTION TO NEURONAL NETWORKS—AND WHEN TO USE THEM

In recent years, deep neural networks (DNNs) have become very popular in machine learning. This led to significant advantages in areas such as perceptual classification, for which classical methods of machine learning had their limitations. Typical examples are image and speech recognition but also text comprehension. The latter makes DNN's particularly interesting for sequence classification as DNA/RNA and proteins can be represented as texts. DNN's were used in several areas of virology, such as the analysis of virus-host protein interaction (Zheng et al., 2019), the detection of novel viruses in human sequencing samples (Tampuu et al., 2019), or the identification of COVID-19 on CT images (Wang et al., 2020).

Neural networks were initially inspired by biological neurons that can perform astonishingly complex tasks thanks to their large number and complex interconnection. Humans can easily recognize connections and patterns, form concepts, and transfer them to new problems. These capabilities are still unmatched in machine learning, but artificial neural networks are candidates to change this. This is partly because neural networks brought a paradigm shift. In classical machine learning, we create a feature set x for our problem of interest and use an algorithm to infer the target value y from these features. This means that with the feature set x, we a priori determine from which properties of the data our algorithm can learn. In particular, the algorithm is limited to these properties. This feature engineering is labor-intensive for humans and limits the algorithm's capacity for machine learning to the user's ability to develop the mentioned features. In contrast, neural networks can identify features in the input data that infer the target value y (see Figure 10.3).

Feature engineering for classic machine learning is a challenging task. This is especially true for biological problems as our knowledge is somewhat limited, and therefore we likely choose features of little biological relevance, which have limited predictive power. In recent years, DNNs that compose numerous layers of neurons have shown their capability to identify even very subtle and complex features. This property is the reason why not only machine learning in general but DNNs, in particular, are of interest for bioinformatics, especially in the context of viruses.

Most DNNs consist of one input layer, several hidden layers, and one output layer. This setup is the architecture of the DNN. Each layer consists of numerous simple, parallel working units, so-called neurons, cells, or nodes. Each node consists of a linear transformation ($z = m \cdot x + n$) and a nonlinear activation function, for example, $\tanh(z)$ (see Figure 10.4a and b). Using nonlinear activation functions, the neural network can represent more complex data and solve more complex tasks.

A neural network processes data by representing them first by numerical values in the input layer. Starting from the input layer, the data is sent along the input layer's edges to other nodes of the network (see Figure 10.5a). The numerical values are changed according to each edge's linear function ($z = m \cdot x + n$). Here, the value m is the weighting of the edge, and the value n is the bias. The changed values then meet new nodes that define their excitation by the activation function (see Figure 10.4). These activation

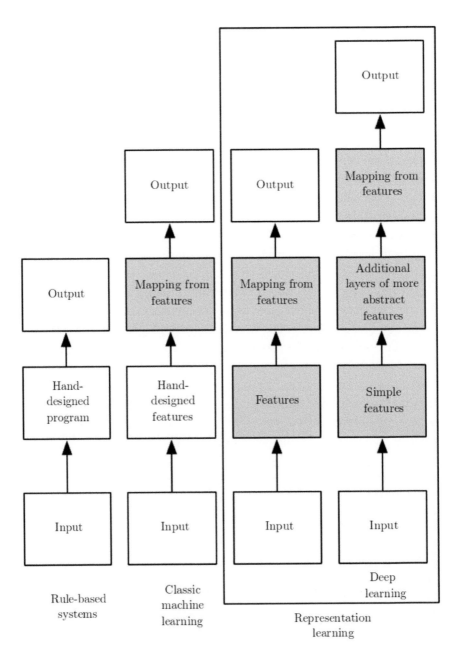

FIGURE 10.3 Comparison between rule-based systems, such as explicitly programmed algorithms that use defined thresholds; classic machine learning approaches, such as SVMs and decision trees that use hand-designed features; and NNs as a part of representation learning, which learn features by itself. Shaded boxes indicate what the algorithms learn from the data. (Provided by Goodfellow et al. (2016).)

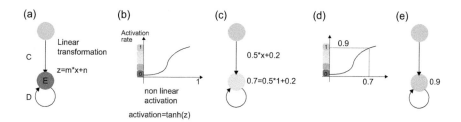

FIGURE 10.4   Working principle of the node activation in a neural network. The general working principle composes of the linear transformation ($z = m \cdot x + n$) of the input signal (a), represented by the vertical edge; the nonlinear activation function (b), represented by the circular edge; and setting of the activation rate, represented by color shift. (c–e) show an example, with weighting $m = 0.5$, a bias $n = 0.2$, and an input signal $x = 1$.

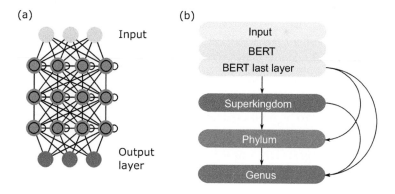

FIGURE 10.5   (a) A simple deep neural network that comprises an input layer, multiple hidden layers, and an output layer. A neural network can contain multiple different types of nodes. (b) A more complex architecture of a taxonomic classification tool (Kretschmer et al., 2021). It consists of a single BERT model (Devlin et al., 2019) that simultaneously predicts all taxonomical ranks (superkingdom, phylum, genus). For this, all output layers (all taxonomical ranks) of the model have access to the BERT model itself and the output layer (the prediction) for higher taxonomical ranks.

values are then sent further through the network until the output layer is reached, and the activations of the output nodes represent the prediction of the neural network. This process transforms the input data into the output data. Before training the neural network, the data transformation happens randomly because all network weights and biases are set randomly. During the training of a neural network, the first performed process is called forward pass. In this step, the network receives input x

and calculates output y'. The next step checks the calculated output and compares it with the expected output y. Next, the backpropagation process starts. In this process, the information from the forward pass is used to gently change the network's parameters so that the next time we use x as input, the calculated y' is closer to y than in the training round before. Therefore, the neural networks learn to predict from input x the output y.

An important advantage of neural networks is that the architecture determines a large part of the predictive power. Therefore, it is common to adopt proven architectures and adapt them to individual needs. For example, the text analysis architecture BERT (Devlin et al., 2019) is used to predict multiple taxonomic classes per sample (Kretschmer et al., 2021) (see Figure 10.5b).

Now that we understand how common machine learning approaches work and understand their importance for modern bioinformatics, we can examine an example. For this, we look at the host prediction of viruses.

10.3 MACHINE LEARNING AS A POWERFUL METHOD TO CLASSIFY VIRAL SEQUENCES

For the problem of sequence classification, mainly supervised machine learning methods are of interest (see Figure 10.1) as, in most cases, we know the classes to be predicted and therefore provide the expected result to the method. Various computational tools for predicting the host of a virus by analyzing its DNA or RNA sequence have been developed. All of these approaches require features to classify the input sequence. To date, it is largely unknown how viruses adapt to new hosts and which mechanisms are responsible for enabling zoonosis (Taubenberger and Kash, 2010; Villordo et al., 2015; Longdon et al., 2014). Because of this incomplete knowledge, it is likely that inappropriate features are selected, *i.e.*, features of little biological relevance, which is problematic for the accuracy of machine learning approaches. In contrast to classic machine learning approaches, deep neural networks can learn features necessary for solving a specific task by themselves.

In the following example, we predict the host of the influenza A virus based on a 400 nt fragment (see Mock et al. (2020) for a detailed description).

A *general workflow* for creating the deep neural networks to predict viral hosts may consist of five major steps (see Figure 10.6a). First, we collect all nucleotide sequences of influenza A viruses with a host label from the European Nucleotide Archive (ENA) database (Leinonen et al., 2010), the ViPR database (Pickett et al., 2012, viprbrc.org/), and the Influenza

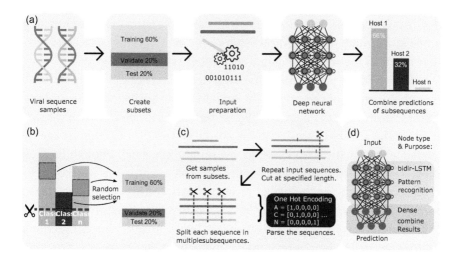

FIGURE 10.6 (a) The general workflow consists of five steps. (b) Balancing of the training set. (c) Conversion of input sequences to gain equal length using the normal repeat method. (d) The bidirectional LSTM analyzes the sequence forward and backward for meaningful patterns, having an awareness of the context as it can remember previously seen data. (Picture adapted from Mock et al. (2020).)

Research Database (National Center for Biotechnology Information, 2017, ncbi.nlm.nih.gov/genomes/FLU/).

We divide the sequences from the selected hosts and viruses into three sets: the training set, the validation set, and the test set in a rather traditional way of 3:1:1. In our example, the number of nucleotide sequences is very heterogeneous for the different hosts, resulting in a biased training. Therefore, we define a fixed validation set and a fixed test set, which consist of an equal number of samples per class (i.e., host). The remaining samples are used to create a variable training set, of which an equal number of samples per class is drawn each training circle (epoch). In the following, we call this the *repeated random undersampling* (see Figure 10.6b).

The training data needs to fulfill several properties to be utilizable for neural networks. The input (being here virus nucleotide sequences) has to be of equal length and also has to be numerical. For the sequence extension, we choose the *normal repeat with gaps* method (see Mock et al. (2020)). This results per virus in sequences of equal length. For each virus, we calculated the length, ranged between 100 and 400 nucleotides, of these subsequences, which results in the least redundant bases. Finally, all subsequences are encoded numerically, using *one hot encoding* to convert the characters A, C, G, T, N into a binary representation (*e.g.*, A = [1,0,0,0,0],

$T = [0,0,0,1,0]$, $N = [0,0,0,0,1]$). Each of these five letters is represented by a one at a unique position/dimension in the vector. Other characters that may occur in the sequence data were treated as the character N.

After the input preparation, the deep neural network predicts the hosts for subsequences of the originally provided viral sequences. The underlying architecture of the neural network dramatically determines its performance. The selected architecture needs to be complex enough to use the available information fully but, at the same time, small enough to avoid memorizing the training set (overfitting). The architecture we choose for the prediction of the influenza A virus hosts consists of three bidirectional LSTM layers (Hochreiter and Schmidhuber, 1997), in the following referred to as *LSTM* architecture (see Figure 10.6d). This bidirectional LSTM tries to find long-term context in the input sequence data, presented to the model in forward and reverse direction (bidirectional), which helps to identify interesting patterns for data classification. The LSTM layers are followed by two dense layers in which the first collects and combines all calculations of the LSTMs, and the second generates the output layer. Each layer consists of 150 nodes with an exception to the output layer, which has a variable number of nodes. Each of the 36 nodes of the output layer represents a possible host species.

In the final step, the predictions of the subsequences are analyzed and combined to a general prediction for their respective original sequences. The neural networks might be trained for 500 epochs during all performed tests. After each epoch, the quality of the model can be evaluated by predicting the hosts of the validation set, comparing the prediction with the true known virus-host pairs. As a metric, we use accuracy. If the current version of the model has a higher validation accuracy than in previous epochs, the model is saved.

10.3.1 Final Host Prediction From Subsequence Predictions

When given a viral nucleotide sequence, the neural network returns the prediction of the host, which is the activation rate of the output nodes. From the output nodes, each node represents a single host. The activation rate of all output nodes adds up to 1.0 and can, therefore, be treated as probabilities. Thus, the activation rate of each output node represents the likelihood of the corresponding species to serve as a host of the given virus sequence. Due to the splitting of the long sequence into multiple subsequences (see Figure 10.6c), the neural network predicts potential hosts for every subsequence. The predictions of the subsequences are then combined to the final prediction of the original sequence. To combine the subsequence prediction into a final sequence prediction, we evaluated the mean.

After combining the subsequence predictions, the single most likely host can be provided as output. However, this limits the prediction power of the neural network. For example, a virus that can survive in two different host species will likely have a high activation score for both hosts. You may also report all possible hosts that reach a certain user-defined likelihood, or it can report the n most likely hosts, where n is also a user-adjustable parameter.

10.4 THE HOST OF A 400 NT FRAGMENT OF INFLUENZA A VIRUS CAN BE PREDICTED VERY ACCURATELY

The influenza A virus dataset has more than 213,000 viral sequences and 36 associated possible host species (32 of them are closely related avian species), a very complex dataset. With 36 different hosts, the expected random accuracy is ~2.78%, which we greatly exceed with over 50% accuracy. Using the top three predictions, we find the correct host in around 76% of the cases.

The deep neural network achieved an AUC of 0.94 (see Figure 10.7)). Despite the close evolutionary distance between the given host species, the trained neuronal network was able to identify potential hosts accurately. We assume that some of the influenza A viruses that are part of the investigated dataset can infect not only one but several host species, *i.e.*, a single viral sequence can occur in more than one host. However, since we only consider a single host species for each tested viral sequence within the test set, the measured accuracy is most likely underestimated. Furthermore, the measured accuracy might be underestimated since most of the misclassifications have been between very closely related hosts.

10.4.1 Varying the Details Yield Similar Predictions

More than 50% accuracy could also be reached with different architecture and voting systems; details can be found in Mock et al. (2020). Overall, the host prediction quality for short subsequences is very high, indicating that an accurate prediction of a viral host is possible even if the given viral sequence is only a fraction of the corresponding genome's size. We additionally tested the CNN+LSTM architectures, being also suitable for host prediction. It is about four times faster and reaches comparable results.

We observed *normal repeat gaps* to be the most suited input preparations for the influenza A virus dataset and the LSTM architecture, as it

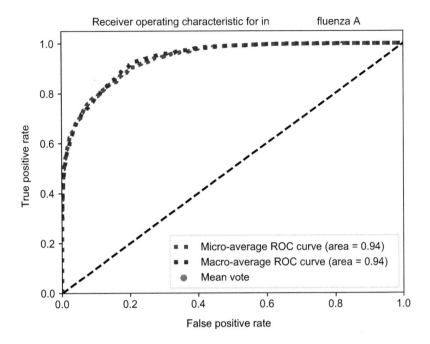

FIGURE 10.7   ROC curve for the influenza A virus dataset, calculated on the test set. The AUC of the micro-average ROC curve is 0.94, and for the macro-average 0.94. The trade-off between false positive rate and true positive rate, when using mean vote, is shown with a green dot.

TABLE 10.1   Percentage Accuracy of the Best Working Input Preparation with Respect to the Combination Methods of the Subsequence Predictions

Combination Method vs Training Setup	Standard	Mean
Influenza A virus		
LSTM, normal repeat gaps	50.14	54.31
CNN+LSTM, random repeat gaps	49.40	52.22
Mean accuracy	49,77	53,27

The combination method with the highest accuracy for each combination is marked in gray.

achieved the highest accuracies. However, datasets and architecture may favor other input preparations.

With the *normal repeat gaps* approach, the neural network can identify the start and the end of the original sequence because gaps mark the ends. This may provide a useful context detection for the LSTM layer.

The presented mean combination of subsequences can increase the accuracy by about 3.5% points (see Table 10.1). Presumably, the prediction combination approach can compensate for the possible information loss caused by the sequence splitting process during the input preparation.

10.5 FINAL REMARKS

In this study, we address the usability of machine learning in virology and demonstrate this using deep learning as an example for predicting hosts for influenza A virus, based solely on viral nucleotide sequences. We established a simple but very capable prediction pipeline, including possible data preparation steps, data training strategies, and a suitable deep neural network architecture.

This pipeline is only an extended example of how to develop a machine learning or, more specifically, a deep learning approach. Thus, this manuscript should be considered more of an introduction and suggestion and is far from complete. Even if one stays with the example of host determination, much can still be illuminated. Be it the lack of generalization across virus species, due to different genome adaptation (Bahir et al., 2009; Mock et al., 2020), or the potential usability of new architectures (Vaswani et al., 2017; Collatz et al., 2020; Kretschmer et al., 2021).

REFERENCES

I. Bahir, M. Fromer, Y. Prat, and M. Linial. Viral adaptation to host: A proeome-based analysis of codon usage and amino acid preferences. *Molecular Systems Biology*, 50 (1):311, 2009.

J. C. Barrett. Monte Carlo simulation of the heterosexual selective spread of the human immunodeficiency virus. *Journal of Medical Virology*, 260 (1):99–109, 1988.

G. Bocharov, N. J. Ford, J. Edwards, T. Breinig, S. Wain-Hobson, and A. Meyerhans. A genetic-algorithm approach to simulating human immunodeficiency virus evolution reveals the strong impact of multiply infected cells and recombination. *Journal of General Virology*, 860 (11):3109–3118, 2005.

M. Collatz, F. Mock, E. Barth, M. Hölzer, K. Sachse, and M. Marz. EpiDope: A deep neural network for linear b-cell epitope prediction. *Bioinformatics*, btaa773, 2020 https://doi.org/10.1093/bioinformatics/btaa773.

W. Commons. File:pca of haplogroup j using 37 strs.png — wikimedia commons, the free media repository, 2020. URL https://commons.wikimedia.org/w/index.php?title=File:PCA_of_Haplogroup_J_using_37_STRs.png&oldid=489805921. [Online; accessed 15-January-2021].

J. Devlin, M.-W. Chang, K. Lee, and K. Toutanova. BERT: Pre-training of deep bidirectional transformers for language understanding. *arXiv*:1810.04805 [cs], May 2019.

M. Ester, H.-P. Kriegel, J. Sander, X. Xu, et al. A density-based algorithm for discovering clusters in large spatial databases with noise. From Kdd-*96 Proceedings*, pp. 226–231, 1996.

J. J. Gart. The analysis of Poisson regression with an application in virology. *Biometrika*, 510 (3/4):517–521, 1964.

R. Gerst and M. Hölzer. PCAGO: An interactive web service to analyze RNA-seq data with principal component analysis. *BioRxiv*, p. 433078, 2018.

H. Gholizade-Narm and A. Noori. Control the population of free viruses in nonlinear uncertain HIV system using q-learning. *International Journal of Machine Learning and Cybernetics*, 90 (7):1169–1179, 2018.

I. Goodfellow, Y. Bengio, and A. Courville. *Deep Learning*, pp. 6–10. MIT Press, Cambridge, MA, 2016. http://www.deeplearningbook.org.

S. Hochreiter and J. Schmidhuber. Long short-term memory. *Neural Computation*, 90 (8):1735–1780, 1997.

Y. Hozumi, R. Wang, C. Yin, and G.-W. Wei. Umap-assisted k-means clustering of large-scale sars-cov-2 mutation datasets. *arXiv preprint*, 2020.

J. Huang, J. Lu, and C. X. Ling. Comparing naive bayes, decision trees, and SVM with AUC and accuracy. *In Third IEEE International Conference on Data Mining*, pp. 553–556. IEEE, Melbourne, FL, 2003.

S. B. Kotsiantis, I. Zaharakis, and P. Pintelas. Supervised machine learning: A review of classification techniques. *Emerging Artificial Intelligence Applications in Computer Engineering*, 1600 (1):3–24, 2007.

F. Kretschmer, F. Mock, A. Kriese, and M. Marz. Bertax, taxonomic classification of DNA sequences with deep neural networks. *BioRxiv*, 2021.

R. Leinonen, R. Akhtar, E. Birney, et al. The European nucleotide archive. *Nucleic Acids Research*, 390 (Suppl 1):D28–D31, 2010.

P. Libin, A. Moonens, T. Verstraeten, F. Perez-Sanjines, N. Hens, P. Lemey, and A. Nowé. Deep reinforcement learning for large-scale epidemic control. *arXiv preprint*, 2020.

J.-Y. Lim, J.-S. Nam, H. Shin, J. Park, H.-I. Song, M. Kang, K.-I. Lim, and Y. Choi. Identification of newly emerging influenza viruses by detecting the virally infected cells based on surface enhanced Raman spectroscopy and principal component analysis. *Analytical Chemistry*, 910 (9):5677–5684, 2019.

B. Longdon, M. A. Brockhurst, C. A. Russell, J. J. Welch, and F. M. Jiggins. The evolution and genetics of virus host shifts. *PLoS Pathogens*, 100 (11):e1004395, 2014.

Y. Lu, J. Liu, C. Lin, H. Wang, Y. Jiang, J. Wang, P. Yang, and F. He. Peroxiredoxin 2: A potential biomarker for early diagnosis of hepatitis b virus related liver fibrosis identified by proteomic analysis of the plasma. *BMC Gastroenterology*, 100 (1):115, 2010.

L. McInnes, J. Healy, and J. Melville. Umap: Uniform manifold approximation and projection for dimension reduction. *arXiv preprint*, 2018.

S. Metzler and O. V. Kalinina. Detection of atypical genes in virus families using a one-class SVM. *BMC Genomics*, 150 (1):913, 2014.

F. Mock, A. Viehweger, E. Barth, and M. Marz. VIDHOP, viral host prediction with deep learning. *Bioinformatics*, btaa705, 2020. https://doi.org/10.1093/bioinformatics/btaa705

S. Modak, S. Mehta, D. Sehgal, and J. Valadi. Application of support vector machines in viral biology. In P. Shapshak, S. Balaji, P. Kangueane, F. Chiappelli, C. Somboonwit, L. J. Menezes, J. T. Sinnott (Eds.), *Global Virology III: Virology in the 21st Century*, pp. 361–403. Springer, Berlin/Heidelberg, 2019.

National Center for Biotechnology Information. Influenza virus database, 2017. [Online; Stand 18. Oktober 2017].

T. T. Nguyen, M. Abdelrazek, D. T. Nguyen, S. Aryal, D. T. Nguyen, and A. Khatami. Origin of novel coronavirus (COVID-19): A computational biology study using artificial intelligence. *BioRxiv*, 2020.

B. E. Pickett, E. L. Sadat, Y. Zhang, et al. ViPR: An open bioinformatics database and analysis resource for virology research. *Nucleic Acids Research*, 400 (D1):D593–D598, 2012.

G. C. Sahoo, M. R. Dikhit, and P. Das. Functional assignment to JEV proteins using SVM. *Bioinformation*, 30 (1):1, 2008.

J. C. Schoeman, J. Hou, A. C. Harms, R. J. Vreeken, R. Berger, T. Hankemeier, and A. Boonstra. Metabolic characterization of the natural progression of chronic hepatitis b. *Genome Medicine*, 80 (1):1–13, 2016.

H. Steinhaus. Sur la division des corp materiels en parties. *Bulletin of the Polish Academy of Sciences* 10 (804):801, 1956.

A. Tampuu, Z. Bzhalava, J. Dillner, and R. Vicente. Viraminer: Deep learning on raw DNA sequences for identifying viral genomes in human samples. *PLoS One*, 140 (9):e0222271, 2019.

S. Tantikitti, S. Tumswadi, and W. Premchaiswadi. Image processing for detection of dengue virus based on WBC classification and decision tree. *In 2015 13th International Conference on ICT and Knowledge Engineering (ICT & Knowledge Engineering 2015)*, pp. 84–89. IEEE, Bangkok, 2015.

J. K. Taubenberger and J. C. Kash. Influenza virus evolution, host adaptation, and pandemic formation. *Cell Host and Microbe*, 70 (6):440–451, 2010.

A. Vaswani, N. Shazeer, N. Parmar, J. Uszkoreit, L. Jones, A. N. Gomez, Ł. Kaiser, and I. Polosukhin. Attention is all you need. *In Advances in Neural Information Processing Systems*, pp. 5998–6008, Long Beach, 2017.

S. M. Villordo, C. V. Filomatori, I. Sánchez-Vargas, C. D. Blair, and A. V. Gamarnik. Dengue virus RNA structure specialization facilitates host adaptation. PLoS Pathog, 11:e1004604, 2015.

S. Wang, B. Kang, J. Ma, et al. A deep learning algorithm using ct images to screen for corona virus disease (COVID-19). MedRxiv, 2020.

S. Wold, K. Esbensen, and P. Geladi. Principal component analysis. *Chemometrics and Intelligent Laboratory Systems*, 20 (1–3):37–52, 1987.

P. Yang, Y. Hwa Yang, B. B. Zhou, and A. Y. Zomaya. A review of ensemble methods in bioinformatics. *Current Bioinformatics*, 50 (4):296–308, 2010.

N. Zheng, K. Wang, W. Zhan, and L. Deng. Targeting virus-host protein interactions: Feature extraction and machine learning approaches. *Current Drug Metabolism*, 200 (3):177–184, 2019.

CHAPTER 11

Master Regulators of Host Response to SARS-CoV-2 as Promising Targets for Drug Repurposing

Manasa Kalya

R&D, geneXplain GmbH
University Medical Center Göttingen

Kamilya Altynbekova

R&D, geneXplain GmbH

Alexander Kel

R&D, geneXplain GmbH
Institute of Chemical Biology and Fundamental Medicine, SBRAS

CONTENTS

11.1 INTRODUCTION

The COVID-19 by SARS-COV-2 belongs to beta-corona virus family, first identified in Wuhan has turned out to be a global pandemic. The primary mode of SARS CoV-2 transmission is through respiratory droplets generated during coughing and sneezing by infected patients.[1] The severity of illness varies from dry cough, fatigue, fever to severe pulmonary damage, and respiratory failure leading to death.[2] Factors like cardiac disease, kidney disease, and age have been associated with poor outcome.[2,3]

COVID-19 like any other RNA viruses first enters the host cell, transcribes its genome, and then translates. The virus has to extensively depend upon the host factors for its replication. The virus has to evade host innate and adaptive immune responses to increase its virulence. This is achieved by modulating host gene expression profiles.[4] On the other side, the response of the host immune system to the infection as well as the severity of the disease strongly depend upon the predisposition of the whole immune system. Earlier researches have reported upregulation of immune activation pathways in peripheral blood,[5] increased mitochondrial activity in bronchoalveolar lavage fluid (BALF),[6] upregulation of pro-viral factors like TYMP, PTGS2, C1S, CFB[7,] and increased cytokine release associated with lung damage[8] that have been revealed based on transcriptome analysis. Therefore, investigating the transcriptomics profiles of the host cell upon infection to SARS-CoV-2 becomes critical in understanding cellular targets of SARS-CoV-2 and underlying host immune parameters. This would pave a way for therapeutic development.

Transcriptomics studies are becoming a standard approach to characterize the pathological state of an affected organism or tissue. Still the challenge remains on how to reveal the underlying molecular mechanisms that render a given pathological state different from the norm. The disease-causing mechanism can be described by a rewiring of the cellular regulatory network, specifically, in the case of infection, the activity of thousands of genes of the host organism is changing in response to the signaling events and regulatory interactions in various cells of the organism. Reconstruction of the disease-specific regulatory networks triggered by the infection can help in identifying! potential master regulators of the respective pathological process. Identification of such master regulators can point to ways on how to block a pathological regulatory cascade. Suppression of certain molecular targets as components of these cascades may stop the pathological process and cure the disease.

Conventional approaches of statistical "-omics" data analysis provide only very limited information about the causes of the observed phenomena and therefore contribute little to the understanding of the pathological molecular mechanism. In contrast, the "upstream analysis" method[9–12] applied here has been devised to provide a casual interpretation of the data obtained for a pathology state. This approach comprises two major steps: (i) analyzing promoters and enhancers of differentially expressed genes for the transcription factors (TFs) involved in their regulation and, thus, important for the process under study; and (ii) reconstructing the signaling pathways that activate these TFs and identifying master regulators at the top of such pathways. For the first step, the database TRANSFAC®[13] is employed together with the TF binding site identification algorithms Match[14] and CMA.[15] The second step involves the signal transduction database TRANSPATH®[16] and special graph search algorithms[10] implemented in the software "Genome Enhancer."

The "upstream analysis" approach has now been extended by a third step that reveals known drugs suitable to inhibit (or activate) the identified molecular targets in the context of the disease under study. This step is performed by using information from HumanPSD™ database.[17] In addition, some known drugs and investigational active chemical compounds are subsequently predicted as potential ligands for the revealed molecular targets. They are predicted using a precomputed database of spectra of biological activities of chemical compounds of a library of 2,507 known drugs and investigational chemical compounds from HumanPSD™ database. The spectra of biological activities for these compounds are computed using the program PASS on the basis of a (Q)SAR approach.[18,19] These predictions can be used for research purposes—for further drug development and drug repurposing initiatives.

During the course of the progression of the COVID-19 disease, the host response evolves gradually leading either to the clearance of the virus or to the development of the severe respiratory syndrome. In our work, we got interested in analyzing gene expression data obtained at different stages of the disease in order to understand the pathological mechanism at the entry place of the virus—in the upper airways and at the place of the severe phase of the disease—in the infected and inflamed lungs. These different states of the disease progression may be characterized by different drug targets and different strategies of the therapy.

Present study is an attempt to conduct transcriptomic analysis of publicly available datasets on COVID-19. We analyzed RNA-seq data from samples of nasopharyngeal (NP) and oropharyngeal (OP) swab of upper airways taken from patients infected by SARS-CoV-2 and also samples from the BALF of hospitalized COVID-19 patients. We identified different pathological mechanisms acting through different sets of master regulators and transcription factors in these tissues that reflect differences in the stages of the disease progression and underlies the necessity of different therapy approaches based on host response in the course of disease progression in COVID-19 infection

11.2 RESULTS

First of all, we analyzed a transcriptomics data set of upper airways samples of COVID-19 patients with acute respiratory illnesses published in ArXive (https://www.ncbi.nlm.nih.gov/pmc/articles/PMC7273244/)—GEO: GSE156063.[20] In the published work, 238 patients with acute respiratory illnesses (ARIs) were tested for SARS-CoV-2 by NP/OP swab PCR and performed host/viral mNGS on the same specimens. The cohort included 94 patients who tested positive for SARS-CoV-2 by PCR, 41 who tested negative but had other pathogenic respiratory viruses detected by mNGS, and 103 with no virus detected (nonviral ARIs). In this study, authors have also measured the virus load in each infected sample by SARS-CoV-2, which gives us possibility to study how response of the cells of the upper airways correlates with the virus load characterizing the very first steps of the disease—first host response upon infection in the upper airways.

In the second data set, we analyzed the host inflammatory response to SARS-CoV-2 infection in infected lungs—at the place of severe stage of the disease. In the recently published paper,[21] authors carried out transcriptome sequencing of the RNAs isolated from the BALF of several hospitalized COVID-19 patients.[21] This analysis gives us a chance to study the pathological mechanism in the lung tissues of the COVID-19 patients at the severe state of the disease.

We performed the analysis of the RNA-seq data using the Genome Enhancer (GE) pipeline that is a fully automated pipeline for omics data analysis, which identifies prospective drug targets and corresponding treatments by reconstructing the molecular mechanism of the studied pathology. In its work, GE uses an upstream analysis algorithm—an integrated promoter and pathway analysis.

We applied GE to the following three datasets: (i) a list of genes with expression values statistically significantly correlated with the virus load in the cells of upper airways (S1); (ii) lists of up- and downregulated genes in the Swab samples of upper airways of the patients with acute respiratory illnesses (ARIs) infected by SARS-CoV-2 virus in comparison with the patients with no virus detected (S2); (iii) lists of up- and downregulated genes in the BALF of hospitalized COVID-19 patients in comparison with gene expression in BALF of healthy individuals (S3).

In the set S1, we identified 260 genes whose expressions are significantly (p-value < 0.05) positively correlated with viral load and 36 genes negatively correlated with the virus load. In order to identify the up- and downregulated genes in the sets S2 and S3, we applied edgeR tool[22] (R/Bioconductor package integrated into the GE pipeline) and compared gene expression in the samples with identified SARS-CoV-2 infection versus samples with no virus detected. The edgeR program calculated the LogFC (the logarithm to the base 2 of the fold change between different conditions), the p-value, and the adjusted p-value (corrected for multiple testing) of the observed fold change. We used a cut-off of adjusted p-value < 0.05 to select the up- and downregulated genes. In Figure 11.1, we show the Venn diagrams comparing the lists of deregulated genes in the studied three sets S1, S2, and S3.

We observed a high heterogeneity of the genes involved in the host response in different tissues and at different states of the host response. Below in Table 11.1, we show the list of top ten genes (out of 77) that are common in all these sets of our study. The full lists of up- and downregulated genes

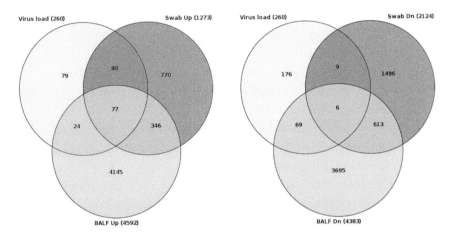

FIGURE 11.1   Venn diagrams showing the distribution of differentially expressed genes found across three datasets.

revealed in our analysis as well as all intermediate results can be found in Genome Enhancer on-line (URL: https://ge.genexplain.com/bioumlweb/#anonymous=true&de=data/Projects/Covid19%20TF/Data/Patient2%20 Keynodes%20for%20best%20model%20viz7%20expr).

Majority of the commonly upregulated genes (51 out of 77) belong to the GO term "response to external biotic stimulus" (p-value<4.6E-32) and 32 genes belong to "defense response to virus" (adj. p-value<2.7E-36). Out of them, 17 genes belong to the Reactome pathway "Interferon Signaling" (adj. p-value<1.1E-24). Interesting is that 14 genes of this set are known "Causal" biomarkers of "Virus diseases" (adj. p-value<5.9E-7), but even more significant, 31 genes are known as "Correlative" biomarkers of "Psoriasis" (p-value<4.5E-16) and "Immune System Diseases" (adj. p-value<1.3E-9).

As it is shown in the Venn diagram, we observed also significant differences in expression patterns of the gene signatures in each condition. These differences may become important for understanding the progression of the pathological processes in COVID-19 and for the identification of optimal therapies.

Let us first consider in detail the S1 set—a set of genes whose expression is correlated with virus load in the cells of upper airways and compare the results of the analysis with other sets.

11.2.1 Functional Classification of Genes

The functional analysis of differentially expressed genes of the S1 set was done by mapping the input 260 genes (that showed the statistically

TABLE 11.1 Table of Top Ten Common Gene Signatures Across Three Datasets Arranged According to LogFC of S2

Symbol	Gene Description	reg_slope (Virus Load)	reg_p_adj (Virus Load)	logFC (Swab Up)	Pvalue (Swab Up)	logFC (BALF Up)	Pvalue (BALF Up)
CXCL11	C-X-C motif chemokine ligand 11	1.658	1.96E-16	5.191	1.57E-45	3.343	1.19E-07
ISG15	ISG15 ubiquitin like modifier	0.978	2.35E-12	3.552	1.65E-37	2.927	1.02E-96
CCL8	C-C motif chemokine ligand 8	1.418	3.57E-12	2.156	3.82E-05	5.920	7.58E-23
PLAAT2	Phospholipase A and acyltransferase 2	0.663	2.14E-08	2.450	3.76E-35	6.873	1.18E-11
CXCL10	C-X-C motif chemokine ligand 10	1.466	7.36E-15	4.095	2.65E-31	1.808	1.18E-04
IDO1	Indoleamine 2,3-dioxygenase 1	0.874	8.82E-11	1.929	7.83E-18	2.783	6.09E-07
IFIT1	Interferon induced protein with tetratricopeptide repeats 1	0.755	1.07E-11	3.129	1.89E-53	1.585	1.38E-15
RSAD2	Radical S-adenosyl methionine domain containing 2	0.616	5.14E-09	2.435	2.91E-30	2.347	4.10E-13
IFIT2	Interferon induced protein with tetratricopeptide repeats 2	1.069	1.08E-13	3.225	6.96E-29	1.110	3.14E-14
ISG20	Interferon stimulated exonuclease gene 20	0.769	1.08E-13	1.478	3.43E-16	3.242	4.24E-29

significant positive correlation with the virus load) to several known ontologies, such as Gene Ontology (GO), disease ontology (based on HumanPSD™ database), and the ontology of signal transduction and metabolic pathways from the *TRANSPATH®* database. Statistical significance was computed using a binomial test. Figures 11.2 and 11.3 show the most significant categories.

So, as we can see from the results of the functional analysis of the genes correlated with virus load in initial stages of infection, the highly dominated response of the host epithelial cells of the upper airways is the immune response, specifically as a response to the viral (or bacterial) invasion. Also, it is clearly seen that many activated genes are known as biomarkers of immune system diseases and biomarkers of response to various RNA viruses. Among the specific activated pathways, we see toll-like receptor pathways, IL-2 pathway and STAT1 pathway.

The common pathways that we identified in all three datasets were "Interferon Signaling" (p-value $< 2.4E-32$ (S1); $1.3E-21$ (S2) and $3.3E-04$ (S3), though showing a gradual decrease of enrichment in these three sets. The specific enriched pathways for the S1 (so, not found in other three datasets) were "Signaling by Interleukins" ($p < 0.002$) and "Cytosolic sensors of pathogen-associated DNA" ($p < 0.0017$) and also pathways "EDAR ---> NF-kappaB" ($p < 0.0028$) and "TLR4 pathway" ($p < 0.009$).

In the set S2, the specific enriched pathways for upregulated genes were "Adaptive Immune System" ($p < 0.0013$) and "Aurora-A cell cycle regulation" ($p < 0.0001$), and for downregulated genes: "Interleukin-1 family signaling" ($p < 3.5E-04$) and "insulin pathway" ($p < 1.3E-04$).

As for the S3 data set, the following specific GO categories and pathways for upregulated genes were identified: "SRP-dependent co-translational protein targeting to membrane" ($p < 8E-28$), "protein targeting to ER" ($p < 1.4E-25$), "Cilium Assembly" ($p < 3.5E-04$), and "Organelle biogenesis and maintenance" ($p < 0.0015$), "Sox9---Smad3---> COL2A1" ($p < 0.0018$), and for the downregulated genes: "Unfolded Protein Response (UPR)" ($p < 1.9E-05$), "Infectious disease" ($p < 5.4E-05$), "Cell-Cell communication" ($p < 0.0046$), "Notch pathway" ($p < 0.012$), and "glycolysis" ($p < 0.021$).

We can see big variability of the pathways that are getting activated at the different stages of the disease. So, this heterogeneity should be taken into account in our efforts of understanding the disease mechanisms and finding the optimal therapy.

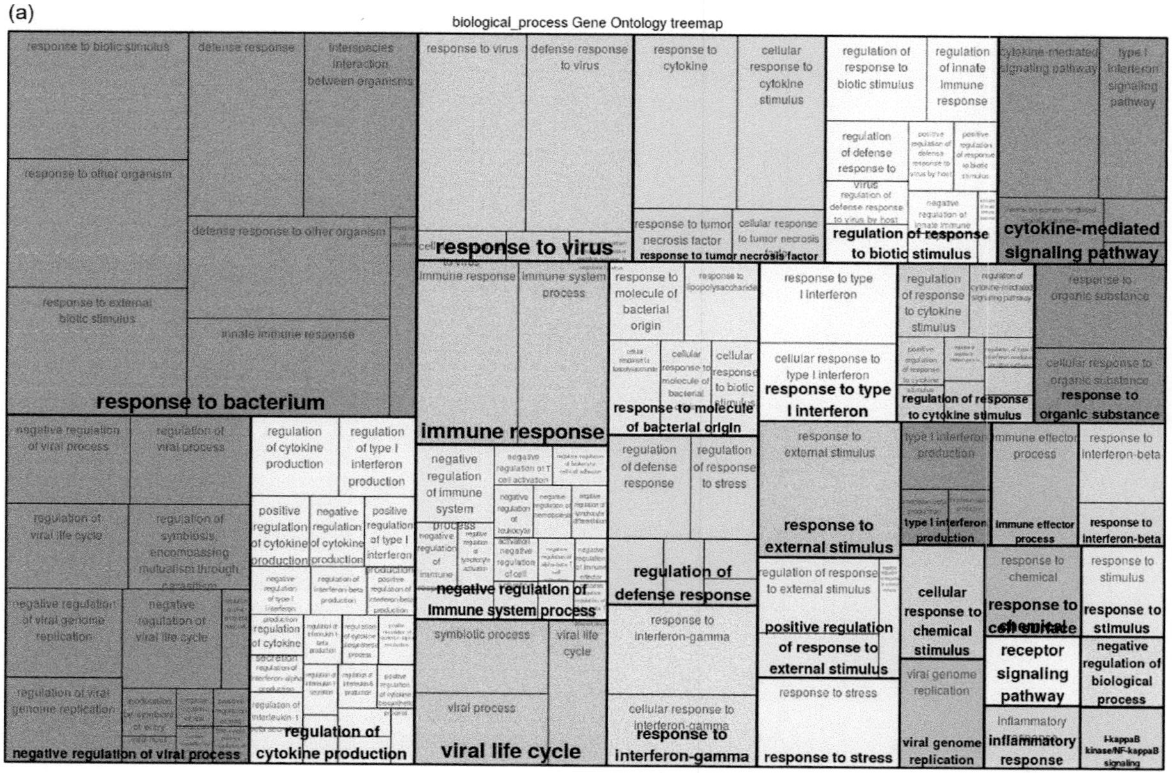

FIGURE 11.2 Results of mapping of genes correlated with virus load to different ontologies: (a) GO (biological processes). The size of each colored box corresponds to the statistical significance of the enrichment of the genes that belong to the category; (b) Enriched TRANSPATH® Pathways. Fifteen top significantly enriched pathways are shown according to the number of genes matches to the category; (c) Enriched HumanPSD™ disease biomarkers. The size of the bars corresponds to the number of biomarkers of the given disease found among the input set.

(Continued)

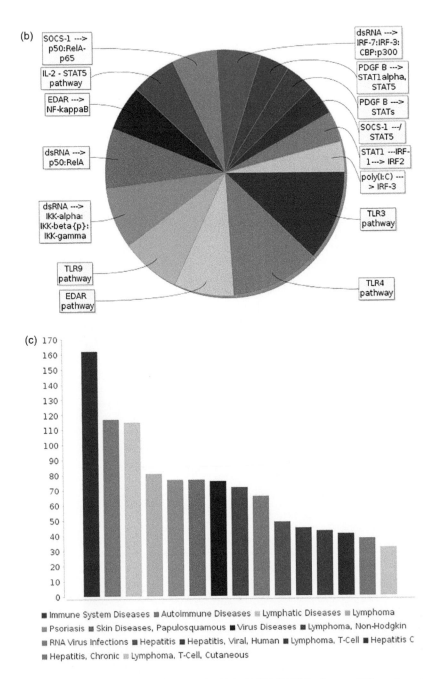

FIGURE 11.2 *(CONTINUED)* (b) Enriched TRANSPATH® Pathways. Fifteen top significantly enriched pathways are shown according to the number of genes matches to the category; (c) Enriched HumanPSD™ disease biomarkers. The size of the bars corresponds to the number of biomarkers of the given disease found among the input set.

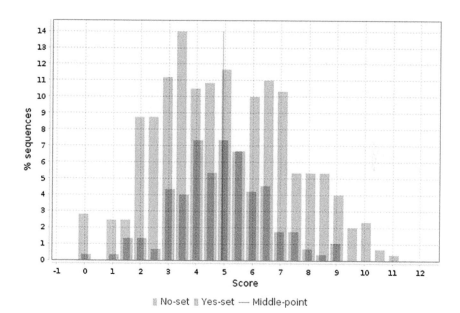

FIGURE 11.3   Comparison of two histograms of the distribution of the composite module score in the promoters of genes correlated with virus load (pink bars) and housekeeping genes (blue bars). Two distributions are statistically different (Wilcoxon p-value (pval): 1.55E-35. The AUC=0.80. The AUC of the model achieves value significantly higher than expected for a random set of regulatory regions (Z-score=3.28).

11.2.2 Analysis of Enriched Transcription Factor Binding Sites and Composite Modules

In the next step, a search for transcription factors binding sites (TFBS) was performed in the regulatory regions of the genes correlating with virus load (target genes) by using the TF binding motif library of the *TRANSFAC®* database. We searched for so-called composite modules that act as potential condition-specific enhancers of the target genes in their upstream regulatory regions (–1,000 bp upstream of transcription start site (TSS)) and identify transcription factors regulating activity of the genes through such enhancers.

Classically, enhancers are defined as regions in the genome that increase transcription of one or several genes when inserted in either orientation at various distances upstream or downstream of the gene.[15] Enhancers typically have a length of several hundreds of nucleotides and are bound by multiple transcription factors in a cooperative manner.[23]

TABLE 11.2   The Optimal Model Found by CMA Algorithm in the
Promoters of Genes Correlating with Virus Load in Upper Airways

Module 1	$W=145$	Module 2	$W=150$
V$IRF_Q6	0.96; $N=2$	V$PR_Q2	0.96; $N=2$
V$POU6F1_01	0.76; $N=3$	V$IRF_Q6	0.96; $N=3$
V$IRF5_Q3	0.78; $N=2$	V$HSF2_Q6	0.89; $N=2$
V$GCM_Q4	0.92; $N=3$	V$SNA_Q6	0.98; $N=2$
V$TCF7RELATED_Q4	0.98; $N=2$	V$STAT3_03	0.82; $N=2$
V$RAD21_10	0.91; $N=3$	V$GKLF_02	0.84; $N=3$
V$STAT5A_03	0.97; $N=3$		

The model contains two modules. Each module characterized by a set of PWMs with their parameters (cut-off, and the maximal number of the considered binding sites N) that identifying TFBS forming clusters in the promoters of the target genes and by its width (W) that reflects the average size of the clusters of sites.

We applied the Composite Module Analyst (CMA)[15] method to detect such potential enhancers, as targets of multiple TFs bound in a cooperative manner to the regulatory regions of the genes of interest. CMA applies a genetic algorithm to construct a generalized model of the enhancers by specifying combinations of TF motifs (from *TRANSFAC®*) whose sites are most frequently clustered together in the regulatory regions of the studied genes. CMA identifies the transcription factors that through their cooperation provide a synergistic effect and thus have a great influence on the gene regulation process.

To build the most specific composite modules, we took promoters of the target genes as the input of CMA algorithm. The obtained CMA model is then applied to compute the CMA score for all target genes.

The model consists of two module(s). Below, in Table 11.2, for each module the following information is shown: PWMs, number of individual matches for each PWM, and score of the best match. The scores of the identified model show a statistically significant difference between our target promoters and promoters of housekeeping genes (Figure 11.3).

We see that identified transcription factors correspond very well to the activated pathways revealed in the previous section. IRF and STAT transcription factors are the end points in the signal transduction pathways enriched by the genes whose expression correlates with virus load.

We compared this result with the promoter analysis results in the datasets S2 and S3. No PWMs for any common TFs were found in the composite models of all three datasets. Composite models of the promoters of the

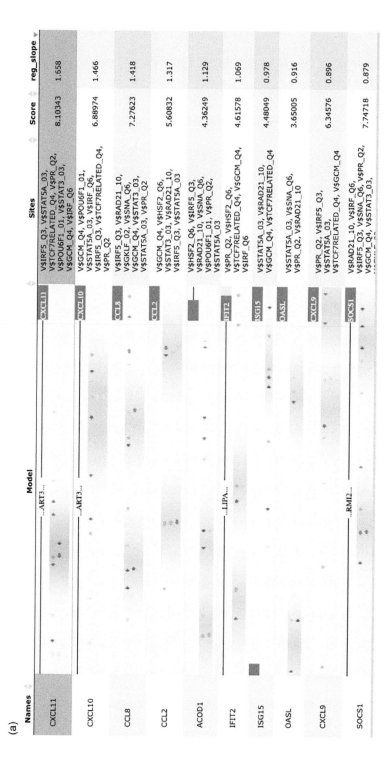

FIGURE 11.4 Maps of TF binding site clusters identified in the promoters of genes correlating with virus load. (a) Maps of top ten genes with maximal correlation with virus load; (b) a detailed map on the chromosomal scale for the promoter of the genes CXCL11. The blue bar corresponds to the first exon of the gene (gene is encoded on the minus strand and goes from right to left). The colored arrows correspond to the TF binding sites identified by the CMA program as a site cluster.

(Continued)

FIGURE 11.4 (*CONTINUED*) Maps of TF binding site clusters identified in the promoters of genes correlating with virus load. (a) Maps of top ten genes with maximal correlation with virus load; (b) a detailed map on the chromosomal scale for the promoter of the genes CXCL11. The blue bar corresponds to the first exon of the gene (gene is encoded on the minus strand and goes from right to left). The colored arrows correspond to the TF binding sites identified by the CMA program as a site cluster.

datasets S1 and upregulated genes of data set S2 have shared PWMs for several IRF factors and for the factor RAD21 that regulates cell survival and G2-M transition. Datasets S1(up) and S3 (up) have common PWMs for two TFs: PR and TCF-7, and datasets S2(up) and S3 (up) have got in common AR transcription factor. Interestingly the data set S3 has got maximal number of specific PWMs for four TFs that were not found in the composite models for S1(up) and S2 (up) promoters. These are the following factors: NF1, POU5F1, MITF, and MAFK. Interestingly that the factor MAFK is known among important antiviral factors induced by type I and type II interferons[24,25] and also involved in blood coagulation. In Table 11.3, we present the full comparison of the PWMs found in the composite models of three datasets.

We analyzed also the downregulated genes in the BALF samples and found several TFs playing an important role in the downregulation of these genes. Among them, only NF1 factor was found playing a role in both up- and downregulated genes. Whereas, we found the following downregulated specific families of factors: Egr, Smad, Gli, and AML.

11.2.3 Finding Master Regulators in Networks

In the second step of the upstream analysis, we identified common regulators of the revealed transcription factors. In total, in the analysis of S1 dataset, we identified 32 genes that encode master regulators in signal transduction network governing the massive genes expression changes in the cells of upper airways in response to the high virus load.

First of all, we ran our key-node analysis algorithm that performs a search for shortest paths through the network of signal transduction reactions (based on TRANSPATH database) upstream of transcription factors identified in the previous step. The maximal depth of the search was 12 steps (taking into account connections between molecular entities and reactions they are involved in). The key-node score was computed for each regulatory molecule in that network that was reachable in 12 steps upstream of the considered TFs. The key-node score reflects the potential of the given molecule to activate the maximal number of the revealed TFs in the shortest "distance" in the considered network. On the next step, we have filtered the key-nodes by the potential presence of a strong positive feedback loop. This was done by checking the criteria that the genes that encode the key-node proteins should belong to the genes whose expression statistically significantly (p-value<0.05) positively correlated with virus

TABLE 11.3 Common and Different Position Weight Matrices (PWMs) of the Composite Models That Form Clusters of TF Sites in the Promoters of Three Analyzed Datasets: S1, S2, and S3

Transpath ID	TF	PWMs of S1 Set (Virus Load)	PWMs of S2 (Up) Set (Swab Up)	PWMs of S3 (Up) Set (BALF Up)
MO000013015	SRF(h)		V$SRF_11	
MO000019320	Bcl-3(h)		V$BCL3_03	
MO000021454	AR(h)		V$AR_14_H	V$AR_Q6
MO000024708	CDP(h)		V$CDP_03	
MO000024750	NF-1C(h)			V$NF1_Q6_02
MO000024986	Myf-6(h)		V$MYF6_06	
MO000025758	MITF(h)			V$MITF_Q6
MO000028668	MafK(h)			V$MAFK_05
MO000028708	NF-1A(h)			V$NF1_Q6_02
MO000028709	NF-1B(h)			V$NF1_Q6_02
MO000056457	TORC2(h)		V$TORC2_Q3	
MO000056618	Oct3(h)			V$POU5F1_01
MO000085551	NF-IX(h)			V$NF1_Q6_02
MO000026306	GCMa(h)	V$GCM_Q4		
MO000161847	GCMb(h)	V$GCM_Q4		
MO000125561	GKLF(h)	V$GKLF_02		
MO000046011	HSF2(h)	V$HSF2_Q6		
MO000031282	IRF-5(h)	V$IRF5_Q3, V$IRF_Q6	V$IRF_Q6	
MO000007686	IRF-1(h)	V$IRF_Q6	V$IRF_Q6	
MO000007691	IRF-2(h)	V$IRF_Q6	V$IRF_Q6	
MO000007703	IRF-7(h)	V$IRF_Q6	V$IRF_Q6	
MO000023424	IRF-8(h)	V$IRF_Q6	V$IRF_Q6	
MO000028705	IRF-4(h)	V$IRF_Q6	V$IRF_Q6	
MO000031288	IRF-6(h)	V$IRF_Q6	V$IRF_Q6	
MO000285816	IRF-3(h)	V$IRF_Q6	V$IRF3_07, V$IRF_Q6	
MO000028320	POU6F1(h)	V$POU6F1_01		
MO000031266	GR(h)	V$PR_Q2		V$GR_Q4
MO000054297	PR(h)	V$PR_Q2		
MO000042938	Rad21(h)	V$RAD21_10	V$RAD21_02	
MO000044348	SNA(h)	V$SNA_Q6		
MO000013123	STAT3(h)	V$STAT3_03		
MO000013125	STAT5A(h)	V$STAT5A_03		
MO000025333	TCF-1(h)	V$TCF7RELATED_Q4		
MO000026845	TCF-3(h)	V$TCF7RELATED_Q4		
MO000159782	LEF-1(h)	V$TCF7RELATED_Q4		V$LEF1_07

load and whose promoters contain clusters of TFBS with the CMA score >0. As the result, we identified 32 such genes.

The final prioritization of the key-nodes was done using two independent criteria (by summing up two ranks). First, the molecules were ranked by the key-node score (applying the cut-off of key-node score >0.2). And, second, the key-nodes were ranked by the CMA score of the promoters of the genes that encode the key-nodes. By summing up these two independent ranks, we computed "Total rank" and prioritized the molecules according to these accumulative rank values. The top key-nodes in this list are the potential master regulators of the pathological process and may be considered as the candidate therapeutic targets as they have a master effect on regulation of the intracellular pathways that are activated in this study. The 32 genes encoding master regulators are shown in Table 11.4.

We performed the master regulator search for the datasets S2 and S3 and compared the results with the results obtained by the analysis of the set S1. The identified master regulators were mapped on the Reactome and TRANSPATH pathways, and the highly enriched pathways were reported. In Figure 11.5, we present the comparison of the enrichment p-values of the revealed pathways in different datasets. One can see that activity of such general pathways of immune response, such as cytokine signaling and signaling by interleukins found quite high at the early stage of infection and becoming less pronounced at the later stage. Whereas the more specific pathways of inflammatory response, such as TNF-alpha pathway, HIF-1 pathway, and TLR3 pathway, are getting highly activated in the BALF samples at the severe stages of the disease. Interesting, the recently created Reactome pathway "SARS-CoV-2 Infection" comprises master regulators that are mainly specific for the severe stage of the COVID-19 disease and only moderately reflect master regulators that are active on the earlier stages. We can also see that pathways such as "cell cycle," "p53 pathway," and "cyclosome regulation" are especially significant in the cellular response of the upper airways and in BALF, but are not really related to the regulation of the genes correlated with the virus load. Another important observation is that the IL-6 pathway, which is considered as one of the most important players in the COVID-19 response to the infection,[26] was identified in our study as significantly enriched in the virus load correlation study only. It was not found highly significant in the samples corresponding to the later stages of the disease.

TABLE 11.4　The 32 Genes Encoding Identified Master Regulators That May Govern the Regulation of the Genes Correlated with Virus Load

Ensembl ID	Gene Symbol	Master Regulator Name	CMA Score	CMA Rank	Keynode Score	Keynode Rank	Total Rank*
ENSG00000158050	DUSP2	PAC-1	10.09	2	0.43	12	14
ENSG00000118503	TNFAIP3	A20	9.74	3	0.46	14	17
ENSG00000137757	CASP5	Caspase-5-isoform1	8.81	5	0.43	15	20
ENSG00000023445	BIRC3	cIAP-2	7.15	19	0.41	23	42
ENSG00000185338	SOCS1	SOCS-1	7.75	10	0.39	35	45
ENSG00000127666	TICAM1	TRIF, dsRNA:TLR3:TRIF	7.37	17	0.47	29	46
ENSG00000124216	SNAI1	SNA	8.04	7	0.34	40	47
ENSG00000136244	IL6	IL-6	7.91	9	0.34	39	48
ENSG00000132274	TRIM22	Staf-50	5.63	34	0.41	17	51
ENSG00000100453	GZMB	GranzymeB(h)	4.67	44	0.48	9	53
ENSG00000184557	SOCS3	SOCS-3	7.51	13	0.40	42	55
ENSG00000138166	DUSP5	DUSP5	3.88	49	0.51	7	56
ENSG00000140464	PML	PML, PML-4	7.42	15	0.43	41	56
ENSG00000140968	IRF8	IRF-8	10.11	1	0.31	55	56
ENSG00000156587	UBE2L6	UbcH8	5.00	40	0.42	16	56
ENSG00000125347	IRF1	IRF-1	8.02	8	0.32	49	57
ENSG00000105639	JAK3	Jak3	5.64	31	0.56	28	59
ENSG00000126561	STAT5A	STAT5A	7.66	12	0.33	47	59

(Continued)

TABLE 11.4 (*Continued*) The 32 Genes Encoding Identified Master Regulators That May Govern the Regulation of the Genes Correlated with Virus Load

Ensembl ID	Gene Symbol	Master Regulator Name	CMA Score	CMA Rank	Keynode Score	Keynode Rank	Total Rank*
ENSG00000124762	CDKN1A	p21Cip1	6.16	30	0.37	32	62
ENSG00000128340	RAC2	Rac2	6.93	21	0.34	43	64
ENSG00000137752	CASP1	Caspase-1, proCaspase-1	3.80	51	0.50	19	70
ENSG00000055332	EIF2AK2	PKR, tr、af6{ub}:TAK1{p}:TAB1{p}:tab 2:PKR	6.71	25	0.39	46	71
ENSG00000101336	HCK	Hck, Hck-p59, Hck-p60	4.94	41	0.53	31	72
ENSG00000120899	PTK2B	Pyk2, Pyk2-isoform1{p}	6.78	22	0.40	53	75
ENSG00000117560	FASLG	(FasL.Fas)6:(Daxx{pS667}:ASK1)2	6.61	29	0.33	48	77
ENSG00000130234	ACE2	ACE2	6.62	28	0.32	50	78
ENSG00000164342	TLR3	TLR3, dsRNA:TLR3:TRIF	7.37	36	0.47	45	81
ENSG00000125826	RBCK1	HOIP:rbck1, sharpin:rbck1:HOIP	5.07	38	0.35	44	82
ENSG00000137193	PIM1	pim1	4.15	46	0.48	37	83
ENSG00000105287	PRKD2	PKD2(h)	5.20	35	0.31	52	87
ENSG00000162772	ATF3	ATF-3	5.09	37	0.31	54	91
ENSG00000232810	TNF	(TNF-alpha:TNFR1) 3:(TRADD:TRAF2) 2:(MEKK1{pT})2, TNF-alpha	3.58	53	0.40	51	104

* Total rank is the sum of the ranks of the master molecules sorted by keynode score and CMA score.

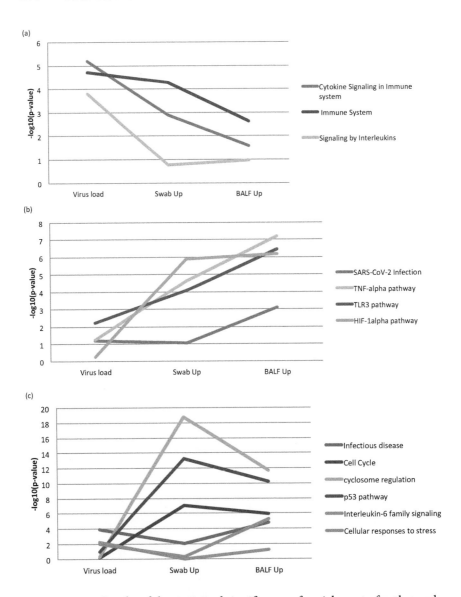

FIGURE 11.5 Graphs of the statistical significance of enrichment of pathways by master regulators identified in different datasets: S1 (Virus load—genes whose expression positively correlates with SARS-CoV-2 virus load), S2 (Swab Up—upregulated genes in the Swab samples of the epithelial cells of upper airways infected by SARS-Cov_2 virus), S3 (BALF Up—upregulated genes in the BALF samples of an infected patient at the late stage of the disease). (a) Pathways with decreasing statistical significance in these three conditions; (b) pathways with increasing statistical significance; (c) pathway with altering statistical significance.

Interestingly, analysis of master regulators controlling downregulation of the genes in BALF dataset revealed TGF-beta pathway as most significant for the downregulation of the genes (p-value<1.0E-16), PRL pathway (p-value<1.5E-13), TL4 pathway (p-value<1.4E-12), IL-3 signaling (p-value<1.5E-10), and several others pathways.

11.2.4 Finding Prospective Drug Targets

Next, the identified master regulators that may govern pathology-associated genes were checked for druggability potential using HumanPSD™[17] database of gene-disease-drug assignments and PASS[19] software for the prediction of biological activities of chemical compounds on the basis of a (Q)SAR approach. Respectively, for each master regulator protein, we have computed two druggability scores: HumanPSD druggability score and PASS druggability score. Where druggability score represents the number of drugs that are potentially suitable for inhibition (or activation) of the corresponding target either according to the information extracted from medical literature (from HumanPSD™ database) or according to cheminformatics predictions of compounds activity against the examined target (from PASS software).

The cheminformatics druggability check is done using a pre-computed database of spectra of biological activities of chemical compounds from a library of all small molecular drugs from HumanPSD™ database, 2507 pharmaceutically active known chemical compounds in total. The spectra of biological activities have been computed using the program PASS on the basis of a (Q)SAR approach.

If both druggability scores were below defined thresholds (see Section 11.4, "Methods," for details), such master regulator proteins were not used in further analysis of drug prediction.

As a result, we created the following two Tables 11.5 and 11.6 of prospective drug targets.

In Figure 11.6, we depicted the intracellular regulatory network controlled by the above-mentioned top master regulators taking into account which of the master regulators are known pharmaceutical targets. This diagram displays the connections between identified transcription factors, which play important roles in the regulation of differentially expressed genes, and top master regulators (pink nodes) with high potential druggability, which are controlling the activity of these TFs (blue nodes). Downstream of the TFs, we show the composite modules identified in the promoters of the target genes (light blue boxes). The positive

TABLE 11.5   Prospective Drug Targets Selected from Full List of Identified Master Regulators Filtered by Druggability Score from HumanPSD™ Database

Gene Symbol	Gene Description	Druggability Score*	Total Rank
IL6	Interleukin 6	8	48
JAK3	Janus kinase 3	3	59
CASP1	Caspase 1	8	70
HCK	HCK proto-oncogene, Src family tyrosine kinase	4	72
PTK2B	Protein tyrosine kinase 2 beta	3	75
ACE2	Angiotensin I converting enzyme 2	4	78

* Druggability score contains the number of drugs that are potentially suitable for inhibition (or activation) of the target. The drug targets are sorted according to the total rank which is the sum of three ranks computed on the basis of the three scores: keynode score, CMA score, and expression change score (logFC, if present). See Section 11.4, "Methods," for details.

TABLE 11.6   Prospective Drug Targets Selected from Full List of Identified Master Regulators Filtered by Druggability Score Predicted by PASS Software

Gene Symbol	Gene Description	Druggability Score*	Total Rank
DUSP2	Dual specificity phosphatase 2	225.15	14
CASP5	Caspase 5	81.32	20
BIRC3	Baculoviral IAP repeat containing 3	497.48	42
SNAI1	Snail family transcriptional repressor 1	189.75	47
IL6	Interleukin 6	799.18	48
TRIM22	Tripartite motif containing 22	482.38	51

* Here, the druggability score for master regulator proteins is computed as a sum of PASS calculated probabilities to be active as a target for various small molecular compounds. The drug targets are sorted according to the Total rank which is the sum of three ranks computed on the basis of the three scores: keynode score, CMA score, and expression change score (logFC, if present). See Section 11.4, "Methods," for details.

feedback loops from the genes to the proteins that are encoded by these genes and playing the master-regulator roles are shown as dotted lines on this diagram.

Below we represent schematically the interactions between the main revealed players of the studied pathology of the response of cells to the virus load. In the schema, we considered the top two drug targets found in this analysis of the gene expression correlated with the virus load. In addition, we have added the top two identified master regulators for which no drugs may be identified yet, but that are playing a crucial role in the molecular mechanism of the studied pathology. Thus we can propose the

(a)

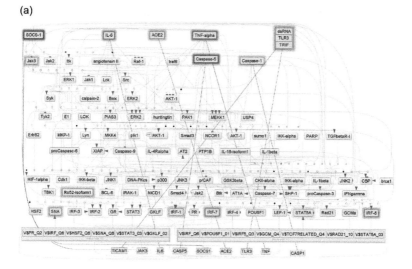

(b)

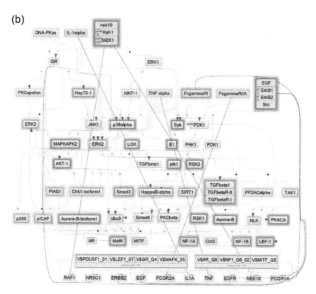

FIGURE 11.6 Diagram of intracellular regulatory signal transduction network identified in the analysis (a) of genes correlated with virus load in the cells of upper airways; (b) of upregulated genes; and (c) of downregulated genes in BALF samples of a severe COVID-19 patient. Master regulators are indicated by red rectangles, transcription factors are blue rectangles, and green rectangles are intermediate molecules, which have been added to the network during the search for master regulators from selected TFs. Orange frames highlight proteins that are encoded by genes whose expression correlates with virus load (shown as light blue boxes at the bottom of the figure). The intensity of the frame color corresponds to the level of the correlation.

(Continued)

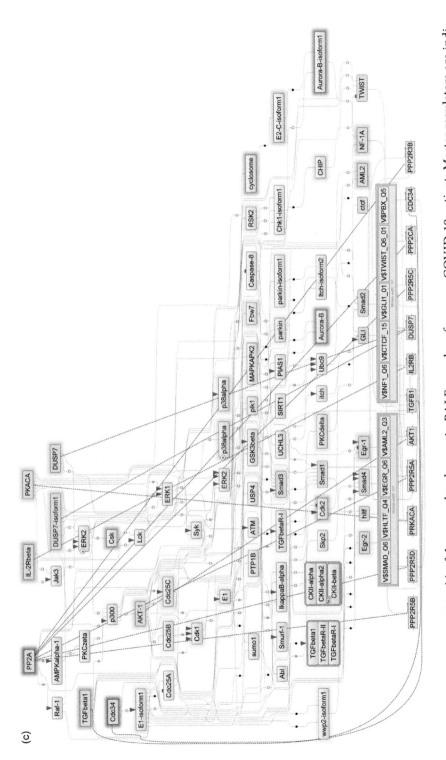

FIGURE 11.6 (*CONTINUED*)   (c) of downregulated genes in BALF samples of a severe COVID-19 patient. Master regulators are indicated by red rectangles, transcription factors are blue rectangles, and green rectangles are intermediate molecules, which have been added to the network during the search for master regulators from selected TFs. Orange frames highlight proteins that are encoded by genes whose expression correlates with virus load (shown as light blue boxes at the bottom of the figure). The intensity of the frame color corresponds to the level of the correlation.

following schema of interactions of the key players of the studied pathology (Figure 11.7).

11.2.5 Identification of Potential Drugs

In the last step of the analysis, we strived to identify known drugs and chemically active compounds that are potentially suitable for inhibition (or activation) of the identified molecular targets in the context of the specified diseases(s). We have specified several diseases of our interest, so drugs that are used to cure such diseases might be useful for the COVID-19 patients. We have included the following diseases into the search for the drugs: coronavirus infections; HIV infections; respiratory distress syndrome; RNA virus infections; severe acute respiratory syndrome. So, in our study, we proposed several top-ranked drug candidates that were found to be active on the identified targets and were selected from four categories:

1. FDA approved drugs or drugs used in clinical trials for the diseases that we have selected;

2. Repurposing drugs used in clinical trials for other diseases;

3. Drugs, predicted by PASS to be active against identified drug targets and against the diseases we have selected;

4. Drugs, predicted by PASS to be active against identified drug targets but for other diseases.

Proposed drugs were selected on the basis of drug rank which was computed from two scores:

- target activity score (depends on ranks of all targets that were found for the selected drug);

- disease activity score (weighted sum of number of clinical trials on the selected diseases where the selected drug is known to be applied) or PASS disease activity score—cheminformatically predicted property of the compound to be active against the selected diseases.

See Section 11.4, "Methods," for more details on drug ranking procedure.
Top drugs of each category are given in Tables 11.7–11.10.

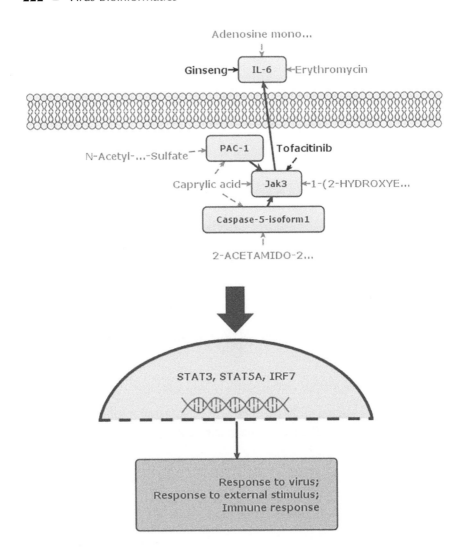

FIGURE 11.7 Schema of interactions of the key players of the studied pathology. Master regulators are shown at the top (green nodes). Arrows between them represent indirect interactions via a cascade of signaling and gene regulatory reactions. Three most prominent TFs identified in the composite modules regulating promoters of the genes whose expression correlates with the SARS-CoV-2 virus load. In violet box, we show the top three GO terms identified for these correlated genes. Drugs and chemical compounds potentially targeting the identified master regulators are shown in this schema. The following drugs and compounds: *Caprylic acid, Adenosine monophosphate, Tofacitinib, 2-acetamido-2-deoxy-beta-d-glucopyranose(beta1–4)-2-acetamido-1,6-anhydro-3-o-[(r)-1-carboxyethyl]-2-deoxy-beta-d-glucopyranose-l-alanyl-gamma-d-glutamyl-meso-diaminopimelyl-d-alanine, Erythromycin, Ginseng,*

11.2.5.1 Drugs Approved in Clinical Trials

TABLE 11.7 FDA Approved Drugs or Drugs Used in Clinical Trials for the Selected Diseases (Coronavirus Infections; HIV Infections; Respiratory Distress Syndrome; RNA Virus Infections; Severe Acute Respiratory Syndrome) These Drugs Are the Most Promising Treatment Candidates Selected for the Identified Drug Targets on the Basis of Literature Curation in HumanPSD™ Database

Drug Name	Target Names	Drug Rank	Disease Activity Score	Phase 4	Status
Lisinopril	ACE2	41	5	HIV infections, arthritis, cardiovascular diseases, deglutition disorders, diabetes mellitus, diabetes mellitus, type 2, diabetic nephropathies…	Small molecule, approved, investigational
Leflunomide	PTK2B	47	2	Arthritis, arthritis, juvenile, arthritis, rheumatoid	Small molecule, approved, investigational
Minocycline	CASP1	50	4	Acne vulgaris, affect, alopecia, autistic disorder, bacterial infections, bipolar disorder, chronic periodontitis…	Small molecule, approved, investigational
IDN-6556	CASP1	55	2	This drug was not tested on Phase 4 clinical trials yet	Small molecule, investigational
Infliximab	TNF	56	5	Arthritis, arthritis, psoriatic, arthritis, rheumatoid, colitis, colitis, ulcerative, Crohn disease, depression…	Biotech, approved

1-(2-hydroxyethyloxymethyl)-6-phenyl thiothymine, and N-Acetyl-D-Galactosamine 6-Sulfate. These drugs and chemical compounds could be considered as a prospective research initiative for drug repurposing and drug development. These drugs were selected as top matching treatments to the most prospective drug targets of the studied pathology; however, these results should be considered with special caution and are to be used for research purposes only, as there is not enough clinical information for adapting these results toward immediate treatment of patients. The drugs given in dark red color on the schema are FDA-approved drugs or drugs which have gone through various phases of clinical trials as active treatments against the selected targets. The drugs given in pink color on the schema are drugs, which were cheminformatically predicted to be active against the selected targets.

TABLE 11.8 Repurposed Drugs Used in Clinical Trials for Other Diseases (Prospective Drugs against the Identified Drug Targets on the Basis of Literature Curation in HumanPSD™ Database)

Drug Name	Target Names	Drug Rank	Phase 4	Status
Ginseng	IL6	19	Fatigue, hematoma, hypersensitivity, infertility, infertility, male	Small molecule, approved, nutraceutical
Tofacitinib	JAK3	24	Arthritis, arthritis, rheumatoid	Small molecule, approved
HMPL-004	TNF, IL6	37	This drug was not tested on Phase 4 clinical trials yet	Small molecule, investigational
Siltuximab	IL6	38	This drug was not tested on Phase 4 clinical trials yet	Biotech, approved

TABLE 11.9 Prospective Drugs, Predicted by PASS Software to Be Active against the Identified Drug Targets with Predicted Activity against the Selected Diseases (Drug Candidates Predicted with the Cheminformatics Tool PASS)

Drug Name	Target Names	Drug Rank	Target Activity Score
Adenosine monophosphate	BIRC3, STAT5A, RBCK1, TNF, IL6, SNAI1, TRIM22…	779	1.19E-2
Adenosine triphosphate	BIRC3, STAT5A, RBCK1, TNF, IL6, SNAI1, TRIM22…	812	1.12E-2
Adenosine-5′-pentaphosphate	BIRC3, STAT5A, RBCK1, TNF, IL6, SNAI1, TRIM22…	812	1.12E-2
1-(2-HYDROXYETHYLOXYMETHYL)-6-PHENYL THIOTHYMINE	PTK2B, EIF2AK2, TNF, HCK, JAK3, IL6	908	9.8E-3
2-ACETAMIDO-2-DEOXY-BETA-D-GLUCOPYRANOSE(BETA1-4)-2-ACETAMIDO-1,6-ANHYDRO-3-O-[(R)-1-CARBOXYETHYL]-2…	BIRC3, IL6, CASP5, CDKN1A	1170	6.33E-3

11.2.5.2 Repurposing Drugs

As the result of drug search, we propose the following drugs as the most promising candidates for treating the pathology under study: lisinopril, ginseng, adenosine monophosphate, and erythromycin. These drugs were selected for acting on the following targets: ACE2, IL6, and BIRC3, which were predicted to be active in the molecular mechanism of the studied pathology.

TABLE 11.10   Prospective Drugs, Predicted by PASS Software to Be Active against the
Identified Drug Targets, though without Cheminformatically Predicted Activity against
the Selected Diseases (Drug Candidates Predicted with the Cheminformatics Tool PASS)

Drug Name	Target Names	Drug Rank	Target Activity Score
Erythromycin	TLR3, TNF, IL6, CDKN1A	192	3.23E-2
Caprylic acid	BIRC3, STAT5A, JAK3, IL6, CASP5, TRIM22, CASP1	205	2.95E-2
Palmitic acid	BIRC3, STAT5A, JAK3, IL6, CASP5, TRIM22, CASP1	205	2.95E-2
Decanoic acid	BIRC3, STAT5A, JAK3, IL6, CASP5, TRIM22, CASP1	205	2.95E-2
Stearic acid	BIRC3, STAT5A, JAK3, IL6, CASP5, TRIM22, CASP1	205	2.95E-2

The selected drugs are top-ranked drug candidates from each of the four categories of drugs: (i) FDA approved drugs or drugs used in clinical trials for the studied pathology; (ii) repurposing drugs used in clinical trials for other pathologies; (iii) drugs, predicted by PASS software to be active against the studied pathology; and (iv) drugs, predicted by PASS software to be repurposed from other pathologies.

The drug candidates revealed here in the analysis of the dataset S1—genes correlated with SARS-CoV-2 virus load in the cells—were compared with the drugs revealed in the analysis of other datasets, specifically with the analysis of data obtained from the BALF samples of severe patients. Finally, two BALF samples of two patients were analyzed with the Genome Enhancer and potential drug candidates were selected and prioritized for these two patients based on the master regulator analysis performed for each sample independently. The results of comparison are shown below. Here, we compared predicted drugs that have been either FDA approved or at least getting through clinical trials of various stages for the list of diseases that we have selected for our analysis.

One can see from the results of this comparison below that: (i) there is a significant number of common drugs identified in these samples. Four drugs were identified as common for all three datasets. These drugs are etanercept (targets: FCGR2A, TNFRSF1B, FCGR1A, TNF), minocycline (targets: VEGFA, CASP1, CASP3, CYCS), thalidomide (targets: FGFR2, TNF), and pseudoephedrine (targets: ADRA2A, ADRB2, TNF). These drugs are currently used or being under clinical trials as a drug against HIV and other infectious diseases and drugs against rheumatoid arthritis, respiratory distress syndrome, and acute lung injury. (ii) Genome

Enhancer has predicted a good number of 18 common drugs for the two analyzed hospitalized COVID-19 patients. The predicted drugs are targeting the following main targets VEGFA, FCGR2A, FCGR1A, IL1R1, IL2RB, EGFR, KIT, and PDGFRA which were found in the analysis of the data of both patients. This demonstrates a significant similarity in activated pathways in the lung tissues of the two infected individuals. (iii) Still there is a clear difference in the lists of predicted drugs for these two patient data. The system has predicted 47 specific drugs that are not predicted for the second patients and also not for the S1 dataset. These drugs affecting the specific targets identified in the data of Patient 1 only, such as IFNAR1, IFNAR2 (predicting the use of several Interferon related drugs), PTPN4, PTPRE (for such drug as Alendronate—known anti-osteoporosis and anti-rheumatoid arthritis drug), GRIN2B, and SLC7A11 (for the drug Acetylcysteine, which is going through clinical trials for respiratory distress Syndrome). Also, for the Patient 2, system has predicted ten unique drugs that are linked to the target NR3C1—the glucocorticoid receptor identified as an important master regulator in Patient 2 (Figure 11.8).

Finally, by summing up several ranks, Genome Enhancer computes a prioritization of the drugs for each analyzed patient data. In Table 11.11, we give top four predicted drugs with the respective drug targets for the two patients analyzed in this study. Using such prioritization, a personalized therapy can be suggested to the patients on the basis of the analysis of each individual transcriptomics profile.

11.3 DISCUSSION

Patients with COVID-19 exhibit a broad spectrum of disease progression, with 81% showing mild, moderate, or no symptoms; 14% showing severe symptoms; and 5% experiencing critical disease with high mortality risk.[27] Age, WBC count, neutrophil count, and fibrinogen values have been reported to significantly vary between mild and severe symptoms.[28] Earlier studies reported that severity and mortality of patients have also been found positively associated with the viral load in the upper airway tract.[29] Based on clinical exhibitions of COVID-19, most of the patients also suffer pneumonia resulting in depleting lung functionality.[30] Gene signatures like virus binding S protein and ACE2 produced in the epithelium of both upper airway and lung were said to play a role in virus replication thereby affecting disease severity.[31,32]

The transcriptome profiling reflects the different reaction of the immune system of different patients leading to different disease progression. In this

BALF Patient 1 (70)

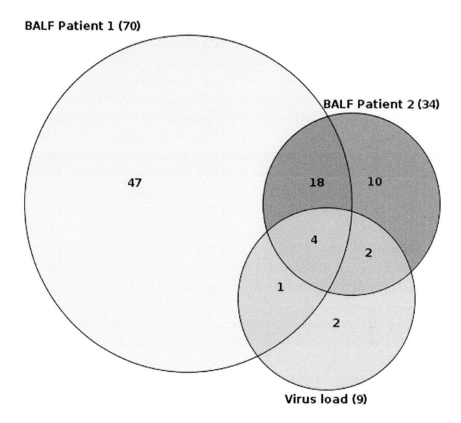

FIGURE 11.8 Comparison of the number of drugs (FDA approved or under clinical trials) predicted in the analysis of the data of the gene expression correlation with the virus load in the epithelial cells of upper airways and in the analysis of upregulated genes in BALF samples of two patients infected by SARS-CoV-2 virus. Patient 2 sample is the S3 dataset which was discussed above. The Patient 1 is a new data set (S4) analyzed here.

chapter, we have performed the computational analysis of SARS-CoV-2 transcriptomics profiles of 238 upper airway tract samples and two BALF samples. Analysis of gene expression changes that correlate with the virus load provides us a clue of the cellular processes and pathways triggered at the early stage of the disease and helped us to identify perspective drug targets and respective drug therapies at this stage. Analysis of the differently expressed genes revealed by transcriptomics profiling of BALF samples allowed us to reveal key pathways activated in the lung tissues at the severe stages of the disease in different patients and helped us to find and prioritize the potential therapies.

We applied the software package "Genome Enhancer" to several datasets of transcriptomics data. The data were preprocessed, statistically analyzed, and differentially expressed genes were identified. Next, we analyzed the enrichment of GO, pathways, and disease categories among the studied gene sets. It clearly shows us signs of activation of various immune-specific pathways, first of all, related to cytokine signaling, as well as regulation of cell proliferation and cell-to-cell communication. Also, we identified a big variability of the activated and inhibited pathways in different datasets that model different stages of the disease. After that, we reconstructed the signal transduction network that regulates differentially expressed genes in the studied stages of the pathology. By analyzing these networks, we have revealed master regulators (signaling proteins and their complexes) that appeared to play a crucial role in the molecular mechanism of the studied pathology. The revealed master regulators are then proposed as the most promising molecular targets for potential drug repurposing and drug development.

Figure 11.9 summarizes important gene regulatory signatures (DEGs, TFs, MRs, and drugs) identified using Genome enhancer pipeline across the datasets.

Let's discuss in detail the identified pathological processes and pathways, revealed master regulators, and proposed potential drug therapies.

With their limited genome size, viruses like SARS-CoV-2 extensively utilize host factors at every step of their life cycle by modulating host cell machinery. Upon infection, viruses induce alterations of intracellular organelles which serve multiple purposes including viral entry, viral gene expression, processing viral proteins, virion structure formation, maturation, and release. In our analysis of BALF samples, we identified that many upregulated proteins are targeting ER and ER-related processes, such as "SRP-dependent co-translational protein targeting to membrane" that is extremely enriched by the upregulated genes (identified as the top revealed GO terms with p-values below 10^{-27}). Extensive growth of ER membrane is required for viral replication and translation and assembly factories, and are most commonly hijacked niche during viral infection. Yet, many aspects of how viruses usurp ER function to promote these different steps are still poorly understood.[33] Broad inhibition of protein translocation to ER is also said to be an interesting target for the development of new anti-cancer drugs as it supports their faster growth.[34]

Another top enriched pathway in our study, revealed in all three datasets analyzed, was the "Interferon Signaling" pathway (with the highest statistical

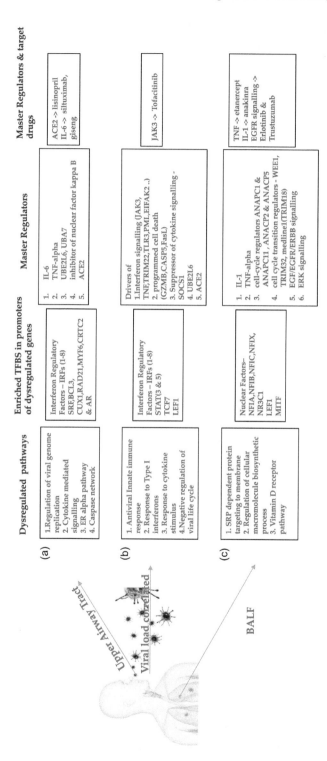

FIGURE 11.9   Important gene regulatory signatures (DEGs, TFs, MRs, and drugs) identified using Genome enhancer pipeline across the datasets. Regulatory signatures of (a) upper airway tract upon infection, (b) correlated with virus load, (c) BALF of infected patients.

significance at early disease state and with lower statistical significance at the late severe state). This finding corresponds to many observations made in the other recent studies, such as the phenomena of downregulation of the type I and type III interferon responses in COVID-19 infected patients.[35] It is hypothesized that the SARS-Cov-2 viral proteins are involved in repressing both the interferon production pathway and the pathways of interferon response.[35,36] In our work, we analyzed promoters of the genes deregulated at the different stages of the COVID-19 disease. We used Composite Module Analyst (CMA) method that helped us to detect potential enhancers as targets of multiple transcription factors bound in a cooperative manner to the regulatory regions of the genes of interest. It is interesting to see that most of these genes have binding sites for and can be potentially regulated by Interferon Regulatory Factors (*IRFs*) (*1–8*), STATs (*3 & 5*), *LEF1*, *TCF7*, and a few other listed above transcription factors. Among them, *IRF1* (significantly upregulated at the early stage, but downregulated at the later severe stage); *IRF3*, *IRF5*, *IRF7*, *IRF8*, and *IRF9* (significantly downregulated at the late severe stage); *STAT1* (significantly upregulated in all stages); and *STAT3*, *STAT5*, and *STAT6* (significantly downregulated in the late severe stage) are not just deregulated but also found as part of the composite elements in the promoters of deregulated genes. Our findings coincide with the conclusions of the impairment of gene regulation by STAT and IRF factors during the SARS-CoV-2 infections.[35] Yet, the opposite behavior is observed between STAT1 and STAT3 as compared to the hypothesis proposed in the cited paper.

One of the most dramatic phenomena observed in many COVID-19 patients is the rapidly developing severe pneumonia symptoms and complications including acute respiratory distress syndrome (ARDS) and multiple organ failure.[37] Clinical features of patients infected with 2019 novel coronavirus in Wuhan, China.[38] It was found that the cascade production of inflammatory cytokines, known as the "cytokine storm," correlates with disease severity during SARS-CoV-2 infection and is characterized by excessive pro-inflammatory cytokine release (*CCL2/MCP-1*, *CXCL10/IP-10*, *CCL3/MIP-1A*, *CXCL1*, *CXCL2*, *CXCL6*, *CXCL8*, *CCL4/MIP-1B*).[21,39] Analysis of the transcriptomics data published in the paper of Xiong et al.[21] allowed us to shed a light on the regulatory mechanism of the initiation of such cytokine storm.

The pathways of activation of various cytokines were identified in all datasets analyzed in our study. As it is seen in Figure 11.5, the general pathway "Cytokine Signaling in Immune system" is activated in all datasets, showing higher significance at the early state of the disease. So, it seems it is specific cytokines rather than the general cytokine pathways that contribute to the severity of the late stages of the disease and the "cytokine storm." It was interesting to observe a clear change of the cytokine repertoire in different datasets analyzed. First of all, the *IL-6*—a pro-inflammatory cytokine that activates *JAK-STAT* signaling—is known to severely aggravate cytokine storm in COVID-19. Ratio of circulating *IL6* and *IFN-Gamma* could be associated with severity in COVID-19 infections.[40] In our analysis, we can see that *IL-6* pathway is enriched by master regulators in the data representing the initial infection and less significant in the BALF tissue of the severe COVID-19 patient (see Figure 11.5). We also noticed that *IFN-gamma* is eight-fold downregulated in BALF sample compared to control, which confirms the earlier observation that it is the ratio between *IL-6* level and *IFN-gamma* level which is important for the development of the severe phenotype rather than the level of the cytokines themselves. It is also proposed earlier that the lower level of the circulating *IFN-gamma* is an important factor for lung fibrosis in COVID-19 patients.[41]

There is a clear correlation between the development of pulmonary fibrosis and respiratory viral infections.[42] Enhanced EGFR signaling (*EGF*, *ERBB2*, *ERBB3*, *ERBB4*, *EGFR*) is also found to promote lung fibrosis upon SARS-CoV infection in mice model[43] and may play role in SARS-CoV-2 replication.[44] Also, it was shown that activation of EGFR by respiratory virus infection (like Influenza virus) suppresses endogenous airway epithelial antiviral signaling.[45] In our analysis, we identified the EGFR complex as one of the important master-regulator in the BALF (Figure 11.6). This confirms that the master-regulator analysis can reveal important factors that are most relevant for the studied pathology.

Another confirmation becomes evident in the results of analysis of downregulated genes in the BALF samples of the COVID-19 patient. In the promoters of the downregulated genes, we revealed composite modules containing sites for the transcription factor Erg-1 (early growth responsive protein-1). It is known that the activity of Egr-1 is directly regulated by *TGF-beta* that stimulates collagen synthesis and accumulation. In our analysis, we identified TGFB1 gene significantly downregulated in the BALF sample (logFC=−3.8). The aberrant TGF-beta signaling is clearly implicated in pathological organ fibrosis[46] and is known to inevitably result in rapid massive fibrosis that remodels and ultimately blocks the airway leading to functional failure of lungs and death.[47] However, the

SARS-CoV-2 receptor and angiotensin II hydrolyzing enzyme (ACE2) are reported to counteract *TGF-beta* mediated collagen deposition in lungs.[47] We observed a high and statistically significant upregulation of ACE2 gene in BALF (logFC=9.4) and a significant positive correlation of expression of ACE2 gene with the virus load in the cells of upper airways.

Also, the transcription factor Egr-1, early growth response protein-1 which gets mainly expressed in response to growth factors or DNA damage has been reported to be playing some role upon viral infections like foot and mouth disease and vaccinia viruses.[48,49] Literature evidence suggest a feedback loop where EGR-1 is induced by *TGF-beta* and mediates the stimulation of collagen gene expression using canonical Smad signaling. Smad4 (found upregulated in our study) is required for normal differentiation of multiple subsets of virus-specific CD8 T cells. In normal circumstances, Smad4 may be activated via a pathway that bypasses the TGF-β receptor.[50] The authors reported that the integrated responses to TGF-β–dependent and TGF-β–independent signaling pathways contribute to the functional heterogeneity of virus-specific memory CD8 T cell populations in the lungs during Influenza infection.[50] *TGF-beta* with its intracellular receptors such as Smads are required for rapid induction of IFN-A and B that are crucial in innate immune responses via Interferon regulatory factors (IRFs). In earlier studies, TGF-β1 and SMAD3 were proposed to be involved in the pathology with HSV-1 infection.[51] Smad3, IFN-A, IFN-E, IFN-L, and IRF6 are significantly highly expressed in our study.

Finally, after revealing the most promising targets, we were able to predict the drug therapies potentially suitable for treating COVID-19 disease at its different stages. In Table 11.7, we present a list of drugs that are predicted on the basis of analysis of the first set of data—the correlation study of the gene expression with virus load in the cells of the upper airways. As we have discussed above, this data models the early stage of the COVID-19 disease. The top predicted drug is *lisinopril*, the specific inhibitor of ACE2 enzyme and receptor, which is known as the receptor for the virus entry into the cells and which was predicted in our analysis as one of the top master regulator of the gene expression correlated with the virus load. The potential role of such inhibitors in COVID-19 therapy is intensively discussed in the recent scientific and medical literature (https://www.clinicaltrialsarena.com/comment/ace-inhibitors-arbs-covid-19/). A clinical trial for testing the therapeutic effect of another ACE inhibitor, *ramipril* is currently ongoing (https://clinicaltrials.gov/ct2/show/NCT04366050) and preliminary observation studies have reported the clinical benefits of it.[52]

As it is currently known, one of the most promising targets for treating COVID-19 diseases is IL-6, which is the most prominent trigger of cytokine storm. More than 60 direct clinical trials are enrolled trying various IL-6 inhibitors in COVID-19 (https://clinicaltrials.gov/ct2/results?cond= &term=tocilizumab+and+covid-19). In our analysis, IL-6 inhibitors are selected by the system as the top potentially repurposing drugs, such as monoclonal antibody *siltuximab* and, most interesting, a drug called *ginseng*, an adaptogen (that increases the body's resistance to stress), which is known in the treatment of viral respiratory infections.[53]

In the course of analysis of BALF transcriptomics from two patients (patient 1 and patient 2), we were able to differentially prioritize drugs for each individual patient (see Table 11.11). Interesting that all proposed drugs in our list are currently considered as promising drugs for COVID-19 treatment. First of all, the corticosteroid *prednisolone* that binds to the glucocorticoid receptor (GR) is in the top four drugs prioritized by our system for patient 1. It is known, that immune cells express receptors for estrogen, androgen, and progesterone and they may modulate immune responses in severe respiratory infections. Certain elegant studies have defined mechanistic roles of these sex hormones and their receptors in innate immune cells to promote resolution of antiviral immunity in lungs.[54] Furthermore, SARS-CoV-2 virus receptor—ACE2—is said to get affected by female hormone.[55] This better explains the fact of GR (glucocorticoid receptor) and AR (androgen receptor) transcription factors emerging in our analysis as important transcription modulators in the BALF dataset. Currently, the corticosteroid *dexamethasone* is the only FDA-approved drug for COVID-19 treatment. It reduces deaths by about 30%, and it is shown that prednisolone may be used if dexamethasone is not available (https://newsnetwork.mayoclinic. org/discussion/covid-19-coronavirus-drugs-are-there-any-that-work/).

Among the other top prioritized drugs for this patient, we identified an anti-arthritis drug *entanercept* that inhibits TNF and FCGR1A and 2A, three master regulators found active in this patient. Several reports are published currently showing positive outcomes of applying this drug to COVID-19 patients.[56] Another anti-arthritis drug, *anakinra* picked up by our analysis is an inhibitor of another master regulator—IL1R1. This drug is often used in cases of sepsis, which is characterized by a hyperinflammatory response and cytokine storm that is observed often in COVID-19 severe cases. The positive outcome of the use of Anakinra in severe COVID-19 patients is recently demonstrated in a small cohort study.[57] Finally, a very interesting direction in the anti-COVID-19 treatment research is repurposing of some

TABLE 11.11 Top Four Drugs Predicted by Genome Enhancer for Two Hospitalized COVID-19 Patients on the Basis of Analysis of Their Transcriptomics Profiles of BALF Samples

Drug	Target Names	Phase 1*	Phase 2	Phase 3	Phase 4	Target Activity Score	Disease Activity Score	Drug Rank
			Patient 1					
Peginterferon alfa-2a	IFNAR1, IFNAR2	Hepatitis, hepatitis B, chronic, hepatitis C, hepatitis C, chronic, hepatitis, chronic	HIV infections, carcinoma, …, hepatitis B,…	HIV infections, …, hepatitis B, … leukemia, …	HIV infections, virus diseases, …, hepatitis B, …, chronic, infection,…	0.877	14	10
Alendronate	PTPN4, PTPRE	… Neoplasms, …, hepatitis B, …	HIV infections, …, arthritis, rheumatoid, asthma,… neoplasms,…	HIV infections,…, arthritis, rheumatoid, asthma,…, neoplasms,…	…, Arthritis, rheumatoid,…, metabolic diseases,…, osteoporosis,…	0.633	5	47
Sunitinib	KIT, PDGFRA	HIV infections, virus diseases, …, neoplasms, …	Virus diseases, …, neoplasms,…	…Neoplasms, carcinoma,…	… Neoplasms,…	0.691	4	48
Bevacizumab	VEGFA, FCGR2A, FCGR1A	… Neoplasms, …	Respiratory distress syndrome, … neoplasms,…	…Neoplasms.…	…Neoplasms,…	0.395	2	46

(*Continued*)

TABLE 11.11 (*Continued*) Top Four Drugs Predicted by Genome Enhancer for Two Hospitalized COVID-19 Patients on the Basis of Analysis of Their Transcriptomics Profiles of BALF Samples

Patient 2

Drug	Target Names	Phase 1*	Phase 2	Phase 3	Phase 4	Target Activity Score	Disease Activity Score	Drug Rank
Etanercept	FCGR2A, TNF, FCGR1A	HIV infections, respiratory distress syndrome,... arthritis, rheumatoid, autoimmune diseases, ...	Respiratory distress syndrome, ..., pneumonia...	..., Arthritis, rheumatoid, ..., pneumonia,...	...Arthritis, rheumatoid... neoplasms	0.356	5	46
Bevacizumab	VEGFA, FCGR2A, FCGR1A	... Neoplasms, ...	Respiratory distress syndrome, ... neoplasms,...	...Neoplasms...	... Neoplasms, ...	0.395	2	46
Anakinra	IL1R1	HIV infections, ..., neoplasms...	...Arthritis, rheumatoid,..., neoplasms,...	...Arthritis, rheumatoid, dermatomyositis, diabetes mellitus, ..., hepatitis, ...	Arthritis, rheumatoid, diabetes mellitus,...	0.407	1	47
Prednisolone	NR3C1	...Arthritis, rheumatoid, asthma, ..., neoplasms, ..., pulmonary disease,...	HIV infections, respiratory distress syndrome, adult, ... virus diseases, ... diabetes mellitus, ...	HIV infections, respiratory distress syndrome...	...Arthritis, rheumatoid, asthma,... diabetes mellitus, ..., pneumonia,..., systemic inflammatory response syndrome,...	0.338	14	49

* In columns Phase 1–Phase 4, we show some main as well as COVID-19 relevant clinical trials for the diseases that the respective drug was going through. Target activity score and disease activity score are computed on the basis of analysis of transcriptomics data and analysis of clinical trial data respectively. Drug rank is computed as a sum of two ranks of Target activity and Disease activity. See Section 11.4, "Methods," for details.

anticancer drugs. Multiple evidence suggest that *bevacizumab*, the inhibitor of Vascular endothelial growth factor (VEGF), is a promising drug for severe and critical COVID-19 patients. VEGF was identified in our analysis as one of the top master-regulator in patient 1 BALF sample, and it is known as the most potent inducing factor to increase vascular permeability, which is often causing acute lung injury in COVID-19 patients. A clinical trial was initiated recently to find out the utility of applying *bevacizumab* to COVID-19 pneumonia (https://clinicaltrials.gov/ct2/show/NCT04305106).

In conclusion, we can say that application of the approach of finding master regulators in the gene regulatory and signal transduction networks helps to decipher the details of the pathological mechanisms on various stages of the COVID-19 disease, identify promising drug targets, and come close to the possibility of personalized medicine by selecting and prioritizing drugs for each individual patient.

11.4 METHODS

11.4.1 Databases Used in the Study

Transcription factor binding sites in promoters and enhancers of differentially expressed genes were analyzed using known DNA-binding motifs described in the *TRANSFAC*® library, release 2020.3 (geneXplain GmbH, Wolfenbüttel, Germany) (*https://genexplain.com/transfac*).[13]

The master regulator search uses the *TRANSPATH*® database[23] (BIOBASE), release 2020.3 (geneXplain GmbH, Wolfenbüttel, Germany) (*https://genexplain.com/transpath*). A comprehensive signal transduction network of human cells is built by the software on the basis of reactions annotated in *TRANSPATH*®.

The information about drugs corresponding to identified drug targets and clinical trials references were extracted from *HumanPSD*™ database,[17] release 2020.3 (*https://genexplain.com/humanpsd*).

The Ensembl database release Human100.38 (hg38)[58] (*http://www.ensembl.org*) was used for gene IDs representation and Gene Ontology (GO)[59] (*http://geneontology.org*) was used for functional classification of the studied gene set.

11.4.2 Methods for the Analysis of Enriched Transcription Factor Binding Sites and Composite Modules

Transcription factor binding sites in promoters and enhancers of differentially expressed genes were analyzed using known DNA-binding motifs.

The motifs are specified using position weight matrices (PWMs) that give weights to each nucleotide in each position of the DNA binding motif for a transcription factor or a group of them.[14]

We search for transcription factor binding sites (TFBS) that are enriched in the promoters and enhancers under study as compared to a background sequence set such as promoters of genes that were not differentially regulated under the condition of the experiment. We denote study and background sets briefly as Yes and No sets. In the current work, we used a workflow considering promoter sequences of a standard length of 1,100 bp (−1,000 to +100). The error rate in this part of the pipeline is controlled by estimating the adjusted p-value (using the Benjamini-Hochberg procedure) in comparison to the TFBS frequency found in randomly selected regions of the human genome (adj. p-value < 0.01).

We have applied the CMA algorithm[15] (Composite Module Analyst) for searching composite modules in the promoters and enhancers of the Yes and No sets. We searched for a composite module consisting of a cluster of 10 TFs in a sliding window of 200–300 bp that statistically significantly separates sequences in the Yes and No sets (minimizing Wilcoxon p-value).

11.4.3 Methods for Finding Master Regulators in Networks

We searched for master regulator molecules in signal transduction pathways upstream of the identified transcription factors. The master regulator search uses a comprehensive signal transduction network of human cells. The main algorithm of the master regulator search has been described earlier.[9] The goal of the algorithm is to find nodes in the global signal transduction network that may potentially regulate the activity of a set of transcription factors found at the previous step of the analysis. Such nodes are considered as the most promising drug targets, since any influence on such a node may switch the transcriptional programs of hundreds of genes that are regulated by the respective TFs.[9,10] In our analysis, we have run the algorithm with a maximum radius of 12 steps upstream of each TF in the input set. The error rate of this algorithm is controlled by applying it 10,000 times to randomly generated sets of input transcription factors of the same set-size. Z-score and FDR value of ranks are calculated then for each potential master regulator node on the basis of such random runs. We control the error rate by the FDR threshold 0.05.

11.4.4 Methods for Analysis of Pharmaceutical Compounds

We seek for the optimal combination of molecular targets (key elements of the regulatory network of the cell) that potentially interact with pharmaceutical compounds from a library of known drugs and biologically active chemical compounds, using information about known drugs from HumanPSD™ and predicting potential drugs using *PASS* program.[19]

11.4.4.1 Method for Analysis of Known Pharmaceutical Compounds

We selected compounds from HumanPSD™ database that have at least one target. Next, we sort compounds using "Drug rank" that is sum of two other ranks:

- ranking by "Target activity score" (T-score$_{PSD}$),

- ranking by "Disease activity score" (D-score$_{PSD}$).

"Target activity score" (T-score$_{PSD}$) is calculated as follows:

$$T - \text{score}_{PSD} = -\frac{|T|}{|T| + w(|AT| - |T|)} \sum_{t \in T} \log_{10}\left(\frac{\text{rank}(t)}{1 + \text{maxRank}(T)}\right),$$

where T is set of all targets related to the compound intersected with input list, $|T|$ is the number of elements in T, AT and $|AT|$ are set of all targets related to the compound and number of elements in it, w is weight multiplier, rank(t) is rank of given target, and maxRank(T) equals max(rank(t)) for all targets t in T.

We use the following formula to calculate "Disease activity score" (D-score$_{PSD}$):

$$D\text{-score}_{PSD} = \begin{cases} \sum_{d \in D} \sum_{p \in P} \text{phase}(d, p) \\ 0, \qquad\qquad D = \varnothing \end{cases},$$

where D is the set of selected diseases, and if D is empty set, D-score$_{PSD} = 0$. P is a set of all known phases for each disease; phase(p, d) equals to the phase number if there are known clinical trials for the selected disease on this phase and zero otherwise.

11.4.4.2 Method for Prediction of Pharmaceutical Compounds

In this study, the focus was put on compounds with high pharmacological efficiency and low toxicity. For this purpose, a comprehensive library of chemical compounds and drugs was subjected to a SAR/QSAR analysis. This library contains 13,040 compounds along with their pre-calculated potential pharmacological activities of those substances, their possible side, and toxic effects, as well as the possible mechanisms of action. All biological activities are expressed as probability values for a substance to exert this activity (Pa).

We selected compounds that satisfied the following conditions:

- Toxicity below a chosen toxicity threshold (defines as Pa, probability to be active as toxic substance).

- For all predicted pharmacological effects that correspond to a set of user-selected disease(s) Pa is greater than a chosen effect threshold.

- There are at least two targets (corresponding to the predicted activity-mechanisms) with predicted Pa greater than a chosen target threshold.

The maximum Pa value for all toxicities corresponding to the given compound is selected as the "Toxicity score." The maximum Pa value for all activities corresponding to the selected diseases for the given compound is used as the "Disease activity score." "Target activity score" (T-score) is calculated as follows:

$$T\text{-score}(s) = -\frac{|T|}{|T| + w(|AT| - |T|)} \sum_{m \in M(s)} Pa(m) \sum_{g \in G(m)} IAP(g)\text{optWeight}(g),$$

where $M(s)$ is the set of activity mechanisms for the given structure (which passed the chosen threshold for activity-mechanisms Pa); $G(m)$ is the set of targets (converted to genes) that corresponds to the given activity-mechanism (m) for the given compound; $Pa(m)$ is the probability to be active of the activity mechanism (m), $IAP(g)$ is the invariant accuracy of prediction for gene from $G(m)$; optWeight(g) is the additional weight multiplier for gene. T is set of all targets related to the compound intersected with input list, $|T|$ is number of elements in T, AT and $|AT|$ are set of all targets related to the compound and number of elements in it, and w is weight multiplier.

"PASS based Disease activity score" (D-score) is calculated as follows:

$$D\text{-score}(g) = IAP(g) \sum_{s \in S(g)} \sum_{m \in M(s,g)} Pa(m),$$

where $S(g)$ is the set of structures for which target list contains given target, $M(s,g)$ is the set of activity mechanisms (for the given structure) that corresponds to the given gene, $Pa(m)$ is the probability to be active of the activity-mechanism (m), $IAP(g)$ is the invariant accuracy of prediction for the given gene.

REFERENCES

1. Mohapatra, S. et al. The novel SARS-CoV-2 pandemic: Possible environmental transmission, detection, persistence and fate during wastewater and water treatment. *Science of the Total Environment* 765, 142746 (2020). doi: 10.1016/j.scitotenv.2020.142746.
2. Rothe, C. et al. Transmission of 2019-nCoV infection from an asymptomatic contact in Germany. *The New England Journal of Medicine* 382, 970–971 (2020).
3. Bermejo-Martin, J. F., Almansa, R., Torres, A., Gonzalez-Rivera, M. & Kelvin, D. J. COVID-19 as a cardiovascular disease: The potential role of chronic endothelial dysfunction. *Cardiovascular Research* 116, E132–E133 (2020).
4. de Wilde, A. H., Snijder, E. J., Kikkert, M. & van Hemert, M. J. Host factors in coronavirus replication. In Arber, W., et al. (eds), *Current Topics in Microbiology and Immunology* vol. 419, pp. 1–42. Springer Verlag: Berlin/Heidelberg (2018).
5. Masood, K. I. et al. Transcriptomic profiling of disease severity in patients with COVID-19 reveals role of blood clotting and vasculature related genes. medRxiv (2020). doi: 10.1101/2020.06.18.20132571.
6. Gardinassi, L. G., Souza, C. O. S., Sales-Campos, H. & Fonseca, S. G. Immune and metabolic signatures of COVID-19 revealed by transcriptomics data reuse. *Frontiers in Immunology* 11, 1636 (2020).
7. Loganathan, T., Ramachandran, S., Shankaran, P., Nagarajan, D. & Suma Mohan, S. Host transcriptome-guided drug repurposing for COVID-19 treatment: A meta-analysis based approach. *PeerJ* 8, e9357 (2020).
8. Tian, S. et al. Pathological study of the 2019 novel coronavirus disease (COVID-19) through postmortem core biopsies. *Modern Pathology* 33, 1007–1014 (2020).
9. Koschmann, J., Bhar, A., Stegmaier, P., Kel, A. & Wingender, E. "Upstream analysis": An integrated promoter-pathway analysis approach to causal interpretation of microarray data. *Microarrays* 4, 270–286 (2015).
10. Boyarskikh, U. et al. Computational master-regulator search reveals mTOR and PI3K pathways responsible for low sensitivity of NCI-H292 and A427 lung cancer cell lines to cytotoxic action of p53 activator Nutlin-3. *BMC Medical Genomics* 11, 12 (2018).
11. Kel, A., Voss, N., Jauregui, R., Kel-Margoulis, O. & Wingender, E. Beyond microarrays: Finding key transcription factors controlling signal transduction pathways. *BMC Bioinformatics* 7, S13 (2006).
12. Stegmaier, P. et al. Advanced computational biology methods identify molecular switches for malignancy in an EGF mouse model of liver cancer. *PLoS One* 6, 17738 (2011).

13. Wingender, E., Dietze, P., Karas, H. & Knüppel, R. TRANSFAC: A database on transcription factors and their DNA binding sites. *Nucleic Acids Research* 24, 238–241 (1996).

14. Kel, A. E. et al. MATCH™: A tool for searching transcription factor binding sites in DNA sequences. *Nucleic Acids Research* 31, 3576–3579 (2003).

15. Waleev, T. et al. Composite module analyst: Identification of transcription factor binding site combinations using genetic algorithm. *Nucleic Acids Research* 34, W541 (2006).

16. Krull, M. et al. TRANSPATH®: An integrated database on signal transduction and a tool for array analysis. *Nucleic Acids Research* 31, 97–100 (2003).

17. Michael, H. et al. Building a knowledge base for systems pathology. *Briefings Bioinformatics* 9, 518–531 (2008).

18. Poroikov, V. V., Filimonov, D. A., Borodina, Y. V., Lagunin, A. A. & Kos, A. Robustness of biological activity spectra predicting by computer program PASS for noncongeneric sets of chemical compounds. *Journal of Chemical Information and Modeling* 40, 1349–1355 (2000).

19. Poroikov, V. V. et al. Computer-aided prediction of biological activity spectra for organic compounds: The possibilities and limitations. *Russian Chemical Bulletin* 68, 2143–2154 (2019).

20. Mick, E. et al. Upper airway gene expression reveals suppressed immune responses to SARS-CoV-2 compared with other respiratory viruses. *Nature Communications* 11, 5854 (2020).

21. Xiong, Y. et al. Transcriptomic characteristics of bronchoalveolar lavage fluid and peripheral blood mononuclear cells in COVID-19 patients. *Emerging Microbes and Infections* 9, 761–770 (2020).

22. McCarthy, D. J., Chen, Y. & Smyth, G. K. Differential expression analysis of multifactor RNA-Seq experiments with respect to biological variation. *Nucleic Acids Research* 40, 4288–4297 (2012).

23. Krull, M. et al. TRANSPATH: An information resource for storing and visualizing signaling pathways and their pathological aberrations. *Nucleic Acids Research* 34, D546–D551 (2006).

24. Liu, S.-Y., Sanchez, D. J., Aliyari, R., Lu, S. & Cheng, G. Systematic identification of type I and type II interferon-induced antiviral factors. *PNAS* 109, 4239–4244 (2012).

25. Kindrachuk, J. et al. Antiviral potential of ERK/MAPK and PI3K/AKT/mTOR signaling modulation for middle east respiratory syndrome coronavirus infection as identified by temporal kinome analysis. *Antimicrobial Agents and Chemotherapy* 59, 1088–1099 (2015).

26. Abbasifard, M. & Khorramdelazad, H. The bio-mission of interleukin-6 in the pathogenesis of COVID-19: A brief look at potential therapeutic tactics. *Life Sciences* 257, 118097 (2020).

27. Tu, Y. F. et al. A review of sars-cov-2 and the ongoing clinical trials. *International Journal of Molecular Sciences* 21, 2657 (2020).

28. Zhou, C. et al. Predictive factors of severe coronavirus disease 2019 in previously healthy young adults: A single-center, retrospective study. *Respiratory Research* 21, 157 (2020).

29. Liu, Y. et al. Viral dynamics in mild and severe cases of COVID-19. *The Lancet Infectious Diseases* 20, 656–657 (2020).

30. Yang, L.-L. & Yang, T. Pulmonary rehabilitation for patients with coronavirus disease 2019 (COVID-19). *Chronic Diseases and Translational Medicine* 6, 79–86 (2020).

31. Lu, R. et al. Genomic characterisation and epidemiology of 2019 novel coronavirus: Implications for virus origins and receptor binding. *Lancet* 395, 565–574 (2020).

32. Liu, M. et al. Potential role of ACE2 in coronavirus disease 2019 (COVID-19) prevention and management. *Journal of Translational Internal Medicine* 8, 9–19 (2020).

33. Romero-Brey, I. & Bartenschlager, R. Endoplasmic reticulum: The favorite intracellular niche for viral replication and assembly. *Viruses* 8, 160 (2016).

34. Van Puyenbroeck, V. & Vermeire, K. Inhibitors of protein translocation across membranes of the secretory pathway: Novel antimicrobial and anticancer agents. *Cellular and Molecular Life Sciences* 75, 1541–1558 (2018).

35. Hadjadj, J. et al. Impaired type I interferon activity and inflammatory responses in severe COVID-19 patients. *Science* 369, 718–724 (2020).

36. Liu, S. Y., Sanchez, D. J., Aliyari, R., Lu, S. & Cheng, G. Systematic identification of type I and type II interferon-induced antiviral factors. *Proceedings of the National Academy of Sciences of the United States of America* 109, 4239–4244 (2012).

37. Huang, Y., Yang, R., Xu, Y. & Gong, P. Clinical characteristics of 36 non-survivors with COVID-19 in Wuhan, China. medRxiv (2020) doi: 10.1101/2020.02.27.20029009.

38. Huang, C. et al. Clinical features of patients infected with 2019 novel coronavirus in Wuhan, China. *Lancet* 395, 497–506 (2020).

39. de la Rica, R., Borges, M. & Gonzalez-Freire, M. COVID-19: In the eye of the cytokine storm. *Frontiers in Immunology* 11, 558898 (2020).

40. Lagunas-Rangel, F. A. & Chávez-Valencia, V. High IL-6/IFN-γ ratio could be associated with severe disease in COVID-19 patients. *Journal of Medical Virology* 92, 1789–1790 (2020).

41. Hu, Z. et al. Clinical characteristics of 24 asymptomatic infections with COVID-19 screened among close contacts in Nanjing, China. *Science China Life Sciences* 63, 706–711 (2020).

42. Jiang, A. G. Viral infection and idiopathic pulmonary fibrosis risk: Still need more evidence. *Chest* 157 1687–1688 (2020).

43. Venkataraman, T. & Frieman, M. B. The role of epidermal growth factor receptor (EGFR) signaling in SARS coronavirus-induced pulmonary fibrosis. *Antiviral Research* 143, 142–150 (2017).

44. Klann, K. et al. Growth factor receptor signaling inhibition prevents SARS-CoV-2 replication. *Molecular Cell* 80, 164–174.e4 (2020).

45. Ueki, I. F. et al. Respiratory virus–induced EGFR activation suppresses IRF1-dependent interferon λ and antiviral defense in airway epithelium. *Journal of Experimental Medicine* 210, 1929–1936 (2013).

46. Sisto, M. et al. The TGF-β1 signaling pathway as an attractive target in the fibrosis pathogenesis of Sjögren's syndrome. *Mediators at Inflammation* 2018, 1–14 (2018).

47. Oliveira, S. C., Delpino, M. V., Giambartolomei, G. H., Quarleri, J. & Splitter, G. Editorial: Advances in liver inflammation and fibrosis due to infectious diseases. *Frontiers Immunology* 11, 1760 (2020).

48. Zhu, Z. et al. Early growth response gene-1 suppresses foot-and-mouth disease virus replication by enhancing type I interferon pathway signal transduction. *Frontiers Microbiology* 9, 2326 (2018).

49. de Oliveira, L. C. et al. The host factor early growth response gene (EGR-1) regulates vaccinia virus infectivity during infection of starved mouse cells. *Viruses* 10, 140 (2018).

50. Hu, Y., Lee, Y.-T., Kaech, S. M., Garvy, B. & Cauley, L. S. Smad4 promotes differentiation of effector and circulating memory CD8 T cells but is dispensable for tissue-resident memory CD8 T cells. *Journal of Immunology* 194, 2407–2414 (2015).

51. Nie, Y. et al. HSV-1 infection suppresses TGF-β1 and SMAD3 expression in human corneal epithelial cells. *Molecular Vision* 14, 1631–1638 (2008).

52. Zhang, P. et al. Association of inpatient use of angiotensin-converting enzyme inhibitors and angiotensin II receptor blockers with mortality among patients with hypertension hospitalized with COVID-19. *Circulation Research* 126, 1671–1681 (2020).

53. Panossian, A. & Brendler, T. The role of adaptogens in prophylaxis and treatment of viral respiratory infections. *Pharmaceuticals* 13, 1–32 (2020).

54. Kadel, S. & Kovats, S. Sex hormones regulate innate immune cells and promote sex differences in respiratory virus infection. *Frontiers in Immunology* 9, 1653 (2018).

55. Liu, J. et al. Sex differences in renal angiotensin converting enzyme 2 (ACE2) activity are 17β-oestradiol-dependent and sex chromosome-independent. *Biology of Sex Differences* 1, 6 (2010).

56. Duret, P.M. Recovery from COVID-19 in a patient with spondyloarthritis treated with TNF-alpha inhibitor etanercept. *Annals of the Rheumatic Diseases.* doi: 10.1136/annrheumdis-2020-217362.

57. Huet, T. et al. Anakinra for severe forms of COVID-19: A cohort study. *Lancet Rheumatology* 2, e393–e400 (2020).

58. Aken, B. L. et al. The Ensembl gene annotation system. *Database* 2016, baw093 (2016).

59. Ashburner, M. et al. Gene ontology: Tool for the unification of biology. *Nature Genetics* 25, 25–29 (2000).

CHAPTER 12

The Potential of Computational Genomics in the Design of Oncolytic Viruses

Henni Zommer and Tamir Tuller

Tel-Aviv University

CONTENTS

12.1 INTRODUCTION

In the age-old search for a cancer cure, there has been one player lurking in the dark, that every few years rears its head and rocks the world into a state of urgency, the virus. What if we can harness these microscopic yet greatly significant viruses to combat diseases that we face?

There is a common hope of finding a cure for cancer when the term cancer in essence is used to describe the common symptoms resulting from different causes [1], which can be considered as different diseases. It is just like taking medicine for a headache can help reduce the discomfort, but the cause of the headache can be many different things, thus not addressing the cause can result in the pain "reappearing" once the effect of the medication wears off [2]. Currently, systemic treatments are used to fight cancer—chemotherapy, radiation therapy, which essentially destroy everything in a nonspecific manner: the healthy, normal cells and tissues alongside the cancer cells. This deems them effective across different cancers, yet they cause great general harm and work productively in only a subset of patients [3]; moreover, recurrent cancer is not uncommon [4]. Such a complex and dynamic "disease" requires therapy that is dynamic and versatile as well. The search for something effective and specific against cancer has been underway for decades, and one field that after years of research is today plausibly the most promising one with regard to cancer-therapy is Oncolytic Virus Therapy (OVT) [5–8]. Using this as part of cancer therapy is believed by many to be the solution for "curing cancer."

For more than a 100 years, viruses have been explored as agents for cancer elimination thanks to case reports from as early as 1904 when a 42-year-old woman with myelogenous leukemia went into remission after a presumed influenza infection (Influenza was identified as a virus more than 30 years later) [9–11]. Later, in 1910, regression of cervical cancer in a patient who received the Pasteur-Roux live attenuated Rabies vaccine was reported [12]. Similar reports were made through the years where the natural viral infection was associated with cancer regression, and interest in the field fluctuated, reaching a peak in the 1950s and 1960s associated with the introduction of rodent models and new methods for virus propagation so that directed evolution was attempted with limited success. This was followed by near-abandonment in the 1970s and 1980s. The development of reverse genetics and genetic engineering had revived interest in OVT in the last two decades, resulting in the first regulatory approval of an oncolytic virus (OV) in 2004, granted in Latvia for the use of Rigvir (Riga virus), an unmodified Enteric Cytopathogenic Human Orphan type 7 (ECHO-7) picornavirus to treat melanoma, which was later approved in Georgia in 2015, and Armenia as of 2016 [13–15]. The first approval of a recombinant OV was given by Chinese regulators for the genetically modified Oncolytic Adenovirus H101 in November 2005 [16,17], followed by FDA approval of Talimogene Laherparepvec (T-VEC) in 2015 [18–20].

Oncolytic viruses are special in that they target tumor cells specifically and have the ability to establish a niche of continuous viral replication within the tumor and tumor environment preferentially resulting in cell

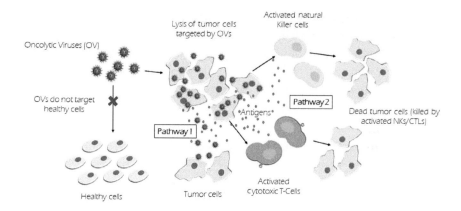

FIGURE 12.1   Oncolytic virus mode of action: 1. By direct lysis of tumor cells and 2. immunogenic response and cell death as a result of release of distress signals/ antigens that result in natural killer (NK) cell activation and killing of tumor cells (innate immunity) and cytotoxic T-cell activation and tumor cell elimination (adaptive immunity).

lysis, while not infecting normal tissue. Though initially the lytic effect was the main attribute associated with OVs, it was later clear that the antitumor effect of OVs acts in an additional manner: aside from direct infection and lysis of tumor cells, OVs arouse the immune system to generate an immune attack that hinders tumor cells (see Figure 12.1) [9,19,21]. Many additional advantages have since then been illuminated with regard to using OVs and how to increase their efficiency [22].

Aside from the natural attributes of OVs that make them advantageous for antitumor therapy, there is the additional advantage that these viruses can be genetically engineered and manipulated to achieve desired characteristics. Furthermore, OVs come in a plethora of shape and forms (several viruses from different families, orders, etc.), so that there is a huge pool to choose from with regard to finding a specific and effective OV for the different types of cancer, stages of cancer, patient genotypes, etc. [21]. This can be the answer not only to cancer therapy, but more specifically to personalized cancer therapy. This packs another punch, since personalized medicine is considered the way of the future [3,23].

There are several challenges in taking this approach from theory to the lab and then to the clinic. In general, a minute portion of what shows promise in the lab translates to the clinic successfully. Having a proper model to work with is a major concern for all studies and is particularly important with OVs since these are dynamic elements that interact with the immune system as well as the tumor microenvironment. The host–pathogen interaction is multidirectional, and thus far we do not have a good model that reflects this accurately. Therefore, this poses a major challenge, since there is a significant gap between the cell and animal models used in the lab and the actual complexity of a human body and the tumor environment [24,25].

Tumor virotherapy will have to be suited per cancer: selection of the optimal virus, since there are many possibilities and each has its own advantages and disadvantages [26]; the administration of the OVT—timing and format of injection/infection; what combinational therapy approach will be used or if single-agent therapy should be used [11], though it seems reasonable that OVT combined with some other therapy will be the answer [25,27]. This emphasizes the challenge of actually testing the different variations of OVT that are suggested—while there are many strategies to optimize OVT, as well as a variety of combinatorial approaches that can be taken, these cannot all be tested and applied to the different types of cancer—that would be too timely, costly, and inefficient. Thus even if an appropriate model was found, only a portion of approaches will be able to be tested, hence it is important to establish a method to select which approach is most promising [9].

In recent years, many excellent reviews and opinions about OVs and OVT have been published [9,12,16,19,21,23,25,27–31]. However, computational biology and genomics in the context of OVs have been largely left out of these. Thus, the aim of this chapter is to discuss the possible (mainly) future contribution of these disciplines to OV design.

Genomic and computational biology, which were found to be helpful in solving challenges such as the development of vaccines and the optimization of heterologous gene expression, can contribute greatly to the challenges in the field by modeling and predicting outcomes, and thus narrowing the scope of the specific approaches that should be prioritized for testing.

12.2 MATHEMATICAL MODELING OF OV

Modern genetic engineering techniques allow for OV optimization in reducing their clinical toxicity, in constructing efficient OV delivery platforms and increasing the efficacy of OVs [10,26], and these have been used for a number of decades thus far [32]. Having quantitative models could greatly reduce time and effort in the search for optimal OVT. This is

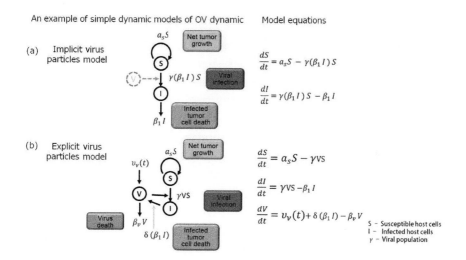

FIGURE 12.2 Simple illustration of mathematical modeling of OV dynamics. The model includes small set differential equations that describe the dynamics of the host cells and the viruses. The variables are defined in the figure and in addition the equations include parameters that should be inferred based on experimental data.

where mathematical modeling can further contribute and has been used to assess oncolytic efficacy [33] based on tumor growth dynamics [34] with Ordinary Differential Equation (ODE) systems typically used for simple models of growth dynamics—quantities of virus over time [35–37].

A simple example of such models appears in Figure 12.2 [10]. The model includes small set differential equations that describe the dynamics of the host cells and the viruses. Typically the variables in this set of equations are the number of (infected) healthy cells, the number of (infected) cancer cells, and the number of viruses.

More complex models such as Partial Differential Equations (PDE) can be used to describe quantities over time and relative dimension in space; this is necessary when the biological effects are predominantly space-dependent (such as diffusion of cancer drugs or viral penetration into a tumor) and cannot be averaged in an ODE model. In addition, Delay Differential Equations (DDE) can be used to mathematically simulate the "eclipse phase" of a virus (the time period from viral entry into a tumor cell to tumor cell burst—lysis). Increasingly complex models can be applied to account for therapeutically relevant parameters such as components of the immune system [10].

A variety of mathematical models have been developed to simulate different immunotherapies [38]. A spatially explicit PDE system, specifically an agent-based model (ABM) system (or a single cell model; each cell is an individual "agent"), may be used to simulate individual cell behaviors and cell–cell or cell–environment interactions that propagate to produce emerging population-level dynamics. From such an approach, one can exclude or identify likely cellular mechanisms that underlie observed complex system-level behavior [39] including that of tumor cells [40–42]. Fast and slow viral growth models are both important to model and account for, since they both occur in a single individual in non-solid and solid tumors respectively; fast growth occurring in little to no restriction, while slow growth occurs in solid tumors, which limit viral growth due to spatial penetration dynamics [43]. There are several examples of such application of modeling [43], with a few cases that validate it with experimental methods [34,44]. Another model was developed to represent syncytia-forming fusion and budding mechanism to lysis based on a PDE model [45]. The ABM model was used by another group to quantify and track spatial patterns, and it was able to qualitatively reproduce the experimental observation and suggests parameters that determine the different growth patterns. The model was further used to study the long-term outcome of the dynamics for the different growth patterns which were supplemented with experimental data [34].

These models can also be used to assess combination approaches [46] as seen in a study to assess the combination of virotherapy with radiation. In one study, they used experimental data they published previously using attenuated MV-NIS (NIS–radiotherapy) for multiple myeloma to infect immunocompromised mice in order to validate an ODE-based mathematical model of radiotherapy [47], and this eventually led to a successful clinical trial [48,49].

The mechanisms and functional dependence of tumor–immune, virus–immune, and virus–cancer interactions are not fully deciphered. Mathematical modeling in close dialog with experimental data may help identify these [50–53]. This could iteratively inform subsequent experimental validation. Collaborations between computational and experimental scientists may be key for enhancing preclinical and eventually clinical efficacies [10].

The main advantage of the mathematical-based modeling of OV is the intuition (at least initial) that they can provide regarding the complex dynamics of OV-cancer-immune system interactions. Their disadvantages include the fact that many times we do not have all the parameters of the models as it is impossible to infer them and/or they are specific to different patients and may change during the "life cycle." Another disadvantage is related to the fact that many times the models are partial and do not include all the many different aspects that can affect the studied phenomena. Specifically, if we want to model the various specific aspects of OVT we should include models with patient specific modeling and accurate parameters related to complex phenomena such as (i) the different interactions of the OV with different subpopulations of cancer and healthy cells; (ii) the three-dimensional diffusion of the OV in the human tissues; (iii) the interactions of the different cells of the immune systems with each other, with the cancer and healthy cells, and with the OV; and (iv) the life cycle of the OV inside the cell. Current models are far from including all these aspects.

12.3 COMPUTATIONAL MODELING OF HETEROLOGOUS GENE EXPRESSION AND LIVE ATTENUATED VACCINES

Two topics are strongly related to computational modeling of OVs, heterologous gene expression, and computational synthetic virology. Thus, in this section, we will briefly discuss these two aspects before discussing the computational design of OV based on genomic data and bioinformatics (the next section).

The modulating of the expression of an endogenous gene of the OV or a transgene introduced into it can significantly affect the characteristics of the OV. Thus, the ability to modulate the expression levels of a gene based on computational models is very relevant to the topic of this review. In recent years, it was demonstrated that the expression levels of heterologous and endogenous genes can be predicted and designed by various computational models. Regression is one method of several, which has been used to demonstrate how features of the transcript can be used to predict the expression levels of genes [54–57]. More sophisticated models that are applied to such research, include biophysical modeling [58,59] and the more recently developed deep neural network model [60]. Such approaches can be used also for designing OVs.

The usage of synthetic and computational virology for designing vaccines is also very relevant to the topic of this review. Computational models have been used successfully in recent years for the generation of viruses and attenuation of viruses [61–67]. Other studies have detected many overlapping codes in the viral coding sequences that are predicted to affect their fitness and modulate viral gene expression [68–74]. Such codes appear in all viruses including those used for OVT and thus need to be considered when designing an OV. Therefore, these and similar approaches can play a significant role in optimizing OV design.

12.4 THE POTENTIAL OF BIOINFORMATICS AND GENOMICS IN THE DEVELOPMENT OF OV

Bioinformatics and genomics studies can haste the process of therapy since they can be integrated into the process of drug discovery—both aiming at search for compounds that are therapeutic as well as finding novel targets [75,76]. These two fields have positively impacted the drug designing process and reduced cost [77] and risk of drug development [78]. Genomics was suggested already two decades ago as having the potential to allow smaller and faster clinical studies, and to individually tailor treatments, with oncology given as an example where the field contributed in such a way [79]. Applying computational methods including data mining semantic methods can advance the development greatly, since a main bottleneck in therapeutics is the validation and advancement in clinical trials. Yet most clinical data is not reported and included in clinical trial information, rather it is medical information collected in the field from patients. Currently there is a higher tendency to save this information in electronic health records; thus, it becomes accessible to informatics approaches, which otherwise, in order to evaluate would require extensive manual work and interpretation. With data-driven computative techniques, this challenge can be overcome and this plethora of information can be included and combined to better characterize and predict side effects, applications, dosages, etc. [80]. The extensive publicly available data sources that include pharmacological, genomic, phenotypic, chemical, and clinical information also make computational drug repositioning possible. Computational tools and algorithms to enable the analysis of such large and complex datasets need to be developed in order to utilize and apply all the information [81,82]. An example of such a case where data-driven drug discovery based on gene expression had contributed in a way that traditional methods would less likely illuminate is a therapeutic association between Cimetidine—a stomach acid reducer that is used to treat and prevent certain types of stomach ulcers as well as gastroesophageal reflux disease and lung adenocarcinoma [83]. Aside from drug repurposing that is generally easier because clinical information with regard to toxicity and dosage exist, these approaches contribute to utilizing metabolites/compounds from different sources for the treatment of human disease, as an example are two cases of plant-based compounds purposed to treat human disease: Gedunin for treatment of prostate cancer [84], and Celastrol to treat acute myeloid leukemia [85]. Furthermore, such approaches can be applied in the clinic to provide noninvasive diagnostic tools [82].

An additional example of where the use of bioinformatics can greatly contribute to current investigations is with regard to the approach of using shRNAs for functional genomics screens for the discovery of clinically relevant cancer gene targets. Algorithms using sequence data and previous experimental results can be used to improve and better direct shRNA design, while better deconvolution methods can shorten the pathway of discovery of tumor-specific targets for improving OVT [86].

Complex biological systems cannot be predicted precisely based only on a small set of equations; thus, indeed some of our best biomedical discoveries were through observation of natural processes, or driving these processes—such as directed evolution, rational design alone may not be efficient and sufficient in bringing us to the goal of a cure [87]. This suggests that we should employ all such directed experimental approaches (including immunology, molecular biology, genomics, structural biology, and so on aside from the obvious field of virology) along with bioinformatics analyses to assess experimental and clinical results and observations so to help explore all possibilities [9] while narrowing our selection as to which approach to advance and examine in the clinic since this is the limiting resource in the search for a cure.

The success thus far of computational biology and genomics in the field of drug discovery and development implies that it can play an integral part in OV design. OVs are actually "smart" drugs which are typically composed of a few macromolecules (RNA/DNA and proteins). These molecules can be programmed to better attack cancer cells and not healthy cells. Tools from the field of computational biology and genomics make it possible to elucidate the actual sequences of nucleotides in the viral genetic material that encode for the desired phenotype of the virus. We believe that the future of OV design will be in general via iterations as illustrated in Figure 12.3: libraries of viruses will be generated using state-of-the-art synthetic biology approaches and will be "brought to life." These viral libraries are then used to infect cells (in vivo and in vitro). The performance of the viruses can be tracked and their effect on cancer cells vs.

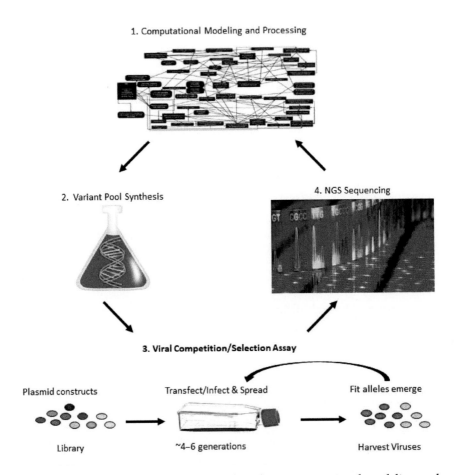

1. Computational Modeling and Processing

2. Variant Pool Synthesis

4. NGS Sequencing

3. Viral Competition/Selection Assay

Plasmid constructs   Transfect/Infect & Spread   Fit alleles emerge

Library   ~4–6 generations   Harvest Viruses

FIGURE 12.3   Illustration of OV design based on computational modeling and genomics. Genomic data is used for developing computational models that then are used for designing libraries of OV (i). State-of-the-art synthetic biology technologies are then used for generating the viruses (ii). Competitions between the viruses are performed in vitro or in vivo (iii) and the performances of the different viruses can be evaluated using NGS (iv). The NGS data can be used for further training and improving the computational models.

healthy cells can be monitored using several methods, including various NGS based experiments; these can include simple sequencing of the viral genomes, yet also more sophisticated protocols for estimating the intracellular gene expression such as RNA-seq and its variations: Ribo-seq [88,89], small -RNA seq [90,91], ChIP-seq [92], etc. (see Ref. [93]). This data is then analyzed using computational and genomic tools, and models can then be used for the design of improved OVs.

12.5 CONCLUSION

There are over 100 clinical trials involving OVs currently, utilizing 13 kinds of OVs, with most being variants of Adenovirus, HSV, and Vaccinia virus [9,26,94]. OVT allows for many variations and approaches to the treatment of cancer, including but not limited to combining the OVs with other forms of known and practiced antitumor therapy (like chemotherapy and radiation therapy), or designing the OVs themselves to express antitumor factors or immune-enhancement factors. The options for treatment with OVs are vast and the plethora of clinical trials exemplify this, as is illustrated by the following examples: A phase I trial of HF10 (Canerpaturev), which is based on an attenuated HSV-1 with a deletion of UL56 and the latency-associated transcript (LAT) [95] that included 28 patients with refractory solid tumors consisting of cutaneous and/or superficial lesions, was completed a few years ago [96] with encouraging results. An example using a combination approach designed within the virus itself is the measles virus engineered to express the sodium iodide symporter gene (MV-NIS), which facilitates the combination of radiovirotherapy to treat taxol- and platinum-resistant ovarian cancer. A clinical evaluation of MV-NIS in patients with this cancer was completed [48], and the results showed that treatment was well tolerated and associated with promising median overall survival. Immune monitoring post-treatment showed treatment triggered cellular immunity against the patients' tumor and suggests the advantage of an added immune mechanism to enhance the antitumor effect—immunovirotherapy [49,97,98].

From these trials, OVT has shown to function well across different cancers, routes of administrations, as monotherapy and in combination therapy. However, OV can be significantly improved to better deal with the challenge of cancer therapy; specifically, today there is no treatment with great success for metastatic cancer. The major challenge related to this objective is the need to deliver the treatment only to the cancer cells and to all of them while dealing with problems such as the immune system, three dimensional diffusion in the body, and the non-homogenous nature of cells (cancerous and healthy). Since viruses have the natural ability to spread and detect/infect specific cells, OVs are natural candidates for solving these challenges.

However, we believe that for the design of OVs to solve such a complicated mission we should include sophisticated modeling and engineering approaches which must be based on accurate measuring of all the parameters of these models. While some simple but very inspiring mathematical models have been suggested, they are far from being suitable for solving the problem. Here we make a stand that the further development of high resolution measurements of OV performance coupled with bioinformatics and genomic approaches can close this gap. We believe that pipelines like the one described in Figure 12.3 can be used in the future for developing efficient OVT for various cancer types. These new tools should of course be integrated with more traditional knowledge and approaches in disciplines such as immunology, oncology, and virology.

REFERENCES

1. Pucci C, Martinelli C, Ciofani G. Innovative approaches for cancer treatment: Current perspectives and new challenges. *Ecancermedical Science.* 2019;13:961. Epub 2019/09/21.
2. Ahmed F. Headache disorders: Differentiating and managing the common subtypes. *British Journal of Pain.* 2012;6(3):124–32. Epub 2012/08/01.
3. Krzyszczyk P, Acevedo A, Davidoff EJ, Timmins LM, Marrero-Berrios I, Patel M, et al. The growing role of precision and personalized medicine for cancer treatment. *Technology (Singap World Sci).* 2018;6(3–4):79–100. Epub 2019/01/11.
4. Ahmed J, Chard LS, Yuan M, Wang J, Howells A, Li Y, et al. A new oncolytic V accinia virus; augments antitumor immune responses to prevent tumor recurrence and metastasis after surgery. *Journal for Immunotherapy of Cancer.* 2020;8(1):e000415.
5. Marelli G, Howells A, Lemoine NR, Wang Y. Oncolytic viral therapy and the immune system: A double-edged sword against cancer. *Frontiers Immunology.* 2018;9:866. Epub 2018/05/15.
6. Vaha-Koskela MJ, Heikkila JE, Hinkkanen AE. Oncolytic viruses in cancer therapy. *Cancer Letters* 2007;254(2):178–216. Epub 2007/03/27.
7. Fukuhara H, Ino Y, Todo T. Oncolytic virus therapy: A new era of cancer treatment at dawn. *Cancer Science.* 2016;107(10):1373–9. Epub 2016/10/30.
8. Howells A, Marelli G, Lemoine NR, Wang Y. Oncolytic viruses-interaction of virus and tumor cells in the battle to eliminate cancer. *Frontiers in Oncology.* 2017;7:195. Epub 2017/09/26.
9. Zheng M, Huang J, Tong A, Yang H. Oncolytic viruses for cancer therapy: Barriers and recent advances. *Molecular Therapy Oncolytics.* 2019;15:234–47. Epub 2019/12/25.
10. Santiago DN, Heidbuechel JPW, Kandell WM, Walker R, Djeu J, Engeland CE, et al. Fighting cancer with mathematics and viruses. *Viruses.* 2017; 9(9):239. Epub 2017/08/24.
11. Eissa IR, Bustos-Villalobos I, Ichinose T, Matsumura S, Naoe Y, Miyajima N, et al. The current status and future prospects of oncolytic viruses in clinical trials against melanoma, glioma, pancreatic, and breast cancers. *Cancers.* 2018;10(10):356. Epub 2018/09/29.

12. Fountzilas C, Patel S, Mahalingam D. Review: Oncolytic virotherapy, updates and future directions. *Oncotarget.* 2017;8(60):102617–39. Epub 2017/12/20.

13. Donina S, Strele I, Proboka G, Auzins J, Alberts P, Jonsson B, et al. Adapted ECHO-7 virus Rigvir immunotherapy (oncolytic virotherapy) prolongs survival in melanoma patients after surgical excision of the tumour in a retrospective study. *Melanoma Research.* 2015;25(5):421–6. Epub 2015/07/21.

14. Babiker HM, Riaz IB, Husnain M, Borad MJ. Erratum: Oncolytic virotherapy including Rigvir and standard therapies in malignant melanoma [Corrigendum]. *Oncolytic Virotherapy.* 2019;8:1. Epub 2019/01/19.

15. Babiker HM, Riaz IB, Husnain M, Borad MJ. Oncolytic virotherapy including Rigvir and standard therapies in malignant melanoma. *Oncolytic Virotherapy.* 2017;6:11–8. Epub 2017/02/23.

16. Kelly E, Russell SJ. History of oncolytic viruses: Genesis to genetic engineering. *Molecular Therapy: The Journal of the American Society of Gene Therapy.* 2007;15(4):651–9. Epub 2007/02/15.

17. Liang M. Oncorine, the world first oncolytic virus medicine and its update in China. *Current Cancer Drug Targets.* 2018;18(2):171–6. Epub 2017/12/01.

18. Ansel A, Rosenzweig JP, Zisman PD, Gesundheit B. Monitoring the efficacy of oncolytic viruses via gene expression. *Frontiers in Oncology.* 2017;7:264. Epub 2017/11/23.

19. Maroun J, Munoz-Alia M, Ammayappan A, Schulze A, Peng KW, Russell S. Designing and building oncolytic viruses. *Future Virology.* 2017;12(4):193–213. Epub 2018/02/02.

20. Andtbacka RH, Kaufman HL, Collichio F, Amatruda T, Senzer N, Chesney J, et al. Talimogene laherparepvec improves durable response rate in patients with advanced melanoma. *Journal of Clinical Oncology* 2015;33(25):2780–8. Epub 2015/05/28.

21. Lawler SE, Speranza MC, Cho CF, Chiocca EA. Oncolytic viruses in cancer treatment: A review. *JAMA Oncology.* 2017;3(6):841–9. Epub 2016/07/22.

22. Mahalingam D, Wilkinson GA, Eng KH, Fields P, Raber P, Moseley JL, et al. Pembrolizumab in combination with the oncolytic virus pelareorep and chemotherapy in patients with advanced pancreatic adenocarcinoma: A phase Ib study. *Clinical Cancer Research: An Official Journal of the American Association for Cancer Research.* 2020;26(1):71–81. Epub 2019/11/07.

23. Raja J, Ludwig JM, Gettinger SN, Schalper KA, Kim HS. Oncolytic virus immunotherapy: Future prospects for oncology. *Journal for Immunotherapy of Cancer.* 2018;6(1):140. Epub 2018/12/06.

24. Wong HH, Lemoine NR, Wang Y. Oncolytic viruses for cancer therapy: Overcoming the obstacles. *Viruses.* 2010;2(1):78–106. Epub 2010/06/15.

25. Ottolino-Perry K, Diallo JS, Lichty BD, Bell JC, McCart JA. Intelligent design: Combination therapy with oncolytic viruses. *Molecular Therapy: The Journal of the American Society of Gene Therapy.* 2010;18(2):251–63. Epub 2009/12/24.

26. Aghi M, Martuza RL. Oncolytic viral therapies: The clinical experience. *Oncogene.* 2005;24(52):7802–16. Epub 2005/11/22.

27. Bommareddy PK, Shettigar M, Kaufman HL. Integrating oncolytic viruses in combination cancer immunotherapy. *Nature Reviews Immunology* 2018;18(8):498–513. Epub 2018/05/11.

28. Bommareddy PK, Kaufman HL. Unleashing the therapeutic potential of oncolytic viruses. *Journal of Clinical Investigation* 2018;128(4):1258–60. Epub 2018/03/06.

29. Kirn D, Martuza RL, Zwiebel J. Replication-selective virotherapy for cancer: Biological principles, risk management and future directions. *Nature Medicine*. 2001;7(7):781–7. Epub 2001/07/04.

30. Chiocca EA, Rabkin SD. Oncolytic viruses and their application to cancer immunotherapy. *Cancer Immunology Research*. 2014;2(4):295–300. Epub 2014/04/26.

31. Chiocca EA. Oncolytic viruses. *Nature Reviews Cancer*. 2002;2(12):938–50. Epub 2002/12/03.

32. Martuza RL, Malick A, Markert JM, Ruffner KL, Coen DM. Experimental therapy of human glioma by means of a genetically engineered virus mutant. *Science*. 1991;252(5007):854–6. Epub 1991/05/10.

33. Biesecker M, Kimn JH, Lu H, Dingli D, Bajzer Z. Optimization of virotherapy for cancer. *Bulletin of Mathematical Biology*. 2010;72(2):469–89. Epub 2009/09/30.

34. Wodarz D, Hofacre A, Lau JW, Sun Z, Fan H, Komarova NL. Complex spatial dynamics of oncolytic viruses in vitro: Mathematical and experimental approaches. *PLoS Computational Biology*. 2012;8(6):e1002547. Epub 2012/06/22.

35. Wodarz D. Computational modeling approaches to studying the dynamics of oncolytic viruses. *Mathematical Biosciences and Engineering: MBE*. 2013;10(3):939–57. Epub 2013/08/03.

36. Murphy H, Jaafari H, Dobrovolny HM. Differences in predictions of ODE models of tumor growth: A cautionary example. *BMC Cancer*. 2016;16:163. Epub 2016/02/28.

37. Gevertz JL, Wares JR. Developing a minimally structured mathematical model of cancer treatment with oncolytic viruses and dendritic cell injections. *Computational and Mathematical Methods in Medicine*. 2018;2018:8760371. Epub 2018/12/05.

38. Wodarz D. Computational modeling approaches to the dynamics of oncolytic viruses. *Wiley Interdisciplinary Reviews Systems Biology and Medicine*. 2016;8(3):242–52. Epub 2016/03/24.

39. Vodovotz Y, An G. Agent-based models of inflammation in translational systems biology: A decade later. *Wiley Interdisciplinary Reviews Systems Biology and Medicine*. 2019;11(6):e1460. Epub 2019/07/02.

40. An G, Mi Q, Dutta-Moscato J, Vodovotz Y. Agent-based models in translational systems biology. *Wiley Interdisciplinary Reviews Systems Biology and Medicine*. 2009;1(2):159–71. Epub 2010/09/14.

41. Wang Z, Butner JD, Kerketta R, Cristini V, Deisboeck TS. Simulating cancer growth with multiscale agent-based modeling. *Seminars in Cancer Biology*. 2015;30:70–8. Epub 2014/05/06.

42. Wang Z, Butner JD, Cristini V, Deisboeck TS. Integrated PK-PD and agent-based modeling in oncology. *Journal of Pharmacokinetics and Pharmacodynamics.* 2015;42(2):179–89. Epub 2015/01/16.

43. Wodarz D, Komarova N. Towards predictive computational models of oncolytic virus therapy: Basis for experimental validation and model selection. *PLoS One.* 2009;4(1):e4271. Epub 2009/01/31.

44. Mok W, Stylianopoulos T, Boucher Y, Jain RK. Mathematical modeling of herpes simplex virus distribution in solid tumors: Implications for cancer gene therapy. *Clinical Cancer Research: An Official Journal of the American Association for Cancer Research.* 2009;15(7):2352–60. Epub 2009/03/26.

45. Jacobsen K, Pilyugin SS. Analysis of a mathematical model for tumor therapy with a fusogenic oncolytic virus. *Mathematical Biosciences.* 2015;270(Pt B):169–82. Epub 2015/03/10.

46. Eftimie R, Hassanein E. Improving cancer detection through combinations of cancer and immune biomarkers: A modelling approach. *Journal of Translational Medicine.* 2018;16(1):73. Epub 2018/03/21.

47. Dingli D, Cascino MD, Josic K, Russell SJ, Bajzer Z. Mathematical modeling of cancer radiovirotherapy. *Mathematical Biosciences.* 2006;199(1):55–78. Epub 2005/12/27.

48. Russell L, Peng KW. The emerging role of oncolytic virus therapy against cancer. *Chinese Clinical Oncology.* 2018;7(2):16. Epub 2018/05/17.

49. Msaouel P, Opyrchal M, Dispenzieri A, Peng KW, Federspiel MJ, Russell SJ, et al. Clinical trials with oncolytic measles virus: Current status and future prospects. *Current Cancer Drug Targets.* 2018;18(2):177–87. Epub 2017/02/24.

50. Rommelfanger DM, Offord CP, Dev J, Bajzer Z, Vile RG, Dingli D. Dynamics of melanoma tumor therapy with vesicular stomatitis virus: Explaining the variability in outcomes using mathematical modeling. *Gene Therapy* 2012;19(5):543–9. Epub 2011/09/16.

51. Eftimie R, Eftimie G. Tumour-associated macrophages and oncolytic virotherapies: A mathematical investigation into a complex dynamics. *Letters in Biomathematics.* 2018;5(supl):S6–S35.

52. Wu JT, Kirn DH, Wein LM. Analysis of a three-way race between tumor growth, a replication-competent virus and an immune response. *Bulletin of Mathematical Biology.* 2004;66(4):605–25. Epub 2004/06/24.

53. Jenner AL, Yun CO, Kim PS, Coster ACF. Mathematical modelling of the interaction between cancer cells and an oncolytic virus: Insights into the effects of treatment protocols. *Bulletin of Mathematical Biology.* 2018; 80(6):1615–29. Epub 2018/04/13.

54. Zur H, Tuller T. Transcript features alone enable accurate prediction and understanding of gene expression in *S. cerevisiae. BMC Bioinformatics.* 2013;14 Suppl 15:S1. Epub 2014/02/26.

55. Ben-Yehezkel T, Atar S, Zur H, Diament A, Goz E, Marx T, et al. Rationally designed, heterologous *S. cerevisiae* transcripts expose novel expression determinants. *RNA Biology.* 2015;12(9):972–84. Epub 2015/07/16.

56. Goodman DB, Church GM, Kosuri S. Causes and effects of N-terminal codon bias in bacterial genes. *Science.* 2013;342(6157):475–9. Epub 2013/09/28.

57. Vogel C, Abreu Rde S, Ko D, Le SY, Shapiro BA, Burns SC, et al. Sequence signatures and mRNA concentration can explain two-thirds of protein abundance variation in a human cell line. *Molecular Systems Biology* 2010;6:400. Epub 2010/08/27.

58. Salis HM, Mirsky EA, Voigt CA. Automated design of synthetic ribosome binding sites to control protein expression. *Nature Biotechnology* 2009; 27(10):946–50. Epub 2009/10/06.

59. Levin D, Tuller T. Whole cell biophysical modeling of codon-tRNA competition reveals novel insights related to translation dynamics. *PLoS Computational Biology.* 2020;16(7):e1008038. Epub 2020/07/11.

60. Tunney R, McGlincy NJ, Graham ME, Naddaf N, Pachter L, Lareau LF. Accurate design of translational output by a neural network model of ribosome distribution. *Nature Structural and Molecular Biology* 2018;25(7):577–82. Epub 2018/07/04.

61. Coleman JR, Papamichail D, Skiena S, Futcher B, Wimmer E, Mueller S. Virus attenuation by genome-scale changes in codon pair bias. *Science.* 2008;320(5884):1784–7. Epub 2008/06/28.

62. Goz E, Tsalenchuck Y, Benaroya RO, Zafrir Z, Atar S, Altman T, et al. Generation and comparative genomics of synthetic dengue viruses. *BMC Bioinformatics.* 2018;19(Suppl 6):140. Epub 2018/05/11.

63. Wimmer E, Mueller S, Tumpey TM, Taubenberger JK. Synthetic viruses: A new opportunity to understand and prevent viral disease. *Nature Biotechnology.* 2009;27(12):1163–72. Epub 2009/12/17.

64. Mueller S, Coleman JR, Papamichail D, Ward CB, Nimnual A, Futcher B, et al. Live attenuated influenza virus vaccines by computer-aided rational design. *Nature Biotechnology* 2010;28(7):723–6. Epub 2010/06/15.

65. Trobaugh DW, Sun C, Dunn MD, Reed DS, Klimstra WB. Rational design of a live-attenuated eastern equine encephalitis virus vaccine through informed mutation of virulence determinants. *PLoS Pathogens.* 2019; 15(2):e1007584. Epub 2019/02/12.

66. Yeh MT, Bujaki E, Dolan PT, Smith M, Wahid R, Konz J, et al. Engineering the live-attenuated polio vaccine to prevent reversion to virulence. *Cell Host and Microbe.* 2020;27(5):736–51 e8. Epub 2020/04/25.

67. Lauring AS, Jones JO, Andino R. Rationalizing the development of live attenuated virus vaccines. *Nature Biotechnology.* 2010;28(6):573–9. Epub 2010/06/10.

68. Goz E, Zafrir Z, Tuller T. Universal evolutionary selection for high dimensional silent patterns of information hidden in the redundancy of viral genetic code. *Bioinformatics.* 2018;34(19):3241–8. Epub 2018/05/03.

69. Zarai Y, Zafrir Z, Siridechadilok B, Suphatrakul A, Roopin M, Julander J, et al. Evolutionary selection against short nucleotide sequences in viruses and their related hosts. *DNA Research: An International Journal for Rapid Publication of Reports on Genes and Genomes.* 2020;27(2): dsaa008.

70. Bahir I, Fromer M, Prat Y, Linial M. Viral adaptation to host: A proteome-based analysis of codon usage and amino acid preferences. *Molcular Systems Biology.* 2009;5:311. Epub 2009/11/06.

71. Aragones L, Guix S, Ribes E, Bosch A, Pinto RM. Fine-tuning translation kinetics selection as the driving force of codon usage bias in the hepatitis A virus capsid. *PLoS Pathogens*. 2010;6(3):e1000797. Epub 2010/03/12.

72. Atkinson NJ, Witteveldt J, Evans DJ, Simmonds P. The influence of CpG and UpA dinucleotide frequencies on RNA virus replication and characterization of the innate cellular pathways underlying virus attenuation and enhanced replication. *Nucleic Acids Research*. 2014;42(7):4527–45. Epub 2014/01/29.

73. Lauring AS, Acevedo A, Cooper SB, Andino R. Codon usage determines the mutational robustness, evolutionary capacity, and virulence of an RNA virus. *Cell Host and Microbe*. 2012;12(5):623–32. Epub 2012/11/20.

74. Holmes EC. The evolutionary genetics of emerging viruses. *Annual Review of Ecology, Evolution, and Systematics*. 2009;40(1):353–72.

75. Florez AF, Park D, Bhak J, Kim BC, Kuchinsky A, Morris JH, et al. Protein network prediction and topological analysis in Leishmania major as a tool for drug target selection. *BMC Bioinformatics*. 2010;11:484. Epub 2010/09/30.

76. Hingorani AD, Kuan V, Finan C, Kruger FA, Gaulton A, Chopade S, et al. Improving the odds of drug development success through human genomics: Modelling study. *Scientific Reports*. 2019;9(1):18911. Epub 2019/12/13.

77. Li GZ, Meng HH, Lu WC, Yang JY, Yang MQ. Asymmetric bagging and feature selection for activities prediction of drug molecules. *BMC Bioinformatics*. 2008;9 (Suppl 6):S7. Epub 2008/06/27.

78. Katara P. Single nucleotide polymorphism and its dynamics for pharmacogenomics. *Interdisciplinary Sciences, Computational Life Sciences*. 2014;6(2):85–92. Epub 2014/08/31.

79. Emilien G, Ponchon M, Caldas C, Isacson O, Maloteaux JM. Impact of genomics on drug discovery and clinical medicine. *QJM: Monthly Journal of the Association of Physicians*. 2000;93(7):391–423. Epub 2000/06/30.

80. Romano JD, Tatonetti NP. Informatics and computational methods in natural product drug discovery: A review and perspectives. *Frontiers in Genetics*. 2019;10:368. Epub 2019/05/23.

81. Fernald GH, Capriotti E, Daneshjou R, Karczewski KJ, Altman RB. Bioinformatics challenges for personalized medicine. *Bioinformatics*. 2011;27(13):1741–8. Epub 2011/05/21.

82. Wooller SK, Benstead-Hume G, Chen X, Ali Y, Pearl FMG. Bioinformatics in translational drug discovery. *Bioscience Reports*. 2017;37(4):BSR20160180. Epub 2017/05/11.

83. Sirota M, Dudley JT, Kim J, Chiang AP, Morgan AA, Sweet-Cordero A, et al. Discovery and preclinical validation of drug indications using compendia of public gene expression data. *Science Translational Medicine* 2011;3(96):96ra77. Epub 2011/08/19.

84. Hieronymus H, Lamb J, Ross KN, Peng XP, Clement C, Rodina A, et al. Gene expression signature-based chemical genomic prediction identifies a novel class of HSP90 pathway modulators. *Cancer Cell*. 2006;10(4):321–30. Epub 2006/10/03.

85. Hassane DC, Guzman ML, Corbett C, Li X, Abboud R, Young F, et al. Discovery of agents that eradicate leukemia stem cells using an in silico

screen of public gene expression data. *Blood.* 2008;111(12):5654–62. Epub 2008/02/29.

86. Mahoney DJ, Stojdl DF. Functional genomic screening to enhance oncolytic virotherapy. *British Journal of Cancer.* 2013;108(2):245–9. Epub 2012/11/22.

87. Bauzon M, Hermiston TW. Oncolytic viruses: The power of directed evolution. *Advances in Virology.* 2012;2012:586389. Epub 2012/02/09.

88. Ingolia NT, Hussmann JA, Weissman JS. Ribosome profiling: Global views of translation. *Cold Spring Harbor Perspectives in Biology.* 2019; 11(5):a032698. Epub 2018/07/25.

89. Lulla V, Dinan AM, Hosmillo M, Chaudhry Y, Sherry L, Irigoyen N, et al. An upstream protein-coding region in enteroviruses modulates virus infection in gut epithelial cells. *Nature Microbiology.* 2019;4(2):280–92. Epub 2018/11/28.

90. Backes S, Langlois RA, Schmid S, Varble A, Shim JV, Sachs D, et al. The Mammalian response to virus infection is independent of small RNA silencing. *Cell Repots.* 2014;8(1):114–25. Epub 2014/06/24.

91. Hess AM, Prasad AN, Ptitsyn A, Ebel GD, Olson KE, Barbacioru C, et al. Small RNA profiling of dengue virus-mosquito interactions implicates the PIWI RNA pathway in anti-viral defense. *BMC Microbiology.* 2011;11:45. Epub 2011/03/02.

92. Gunther T, Theiss JM, Fischer N, Grundhoff A. Investigation of viral and host chromatin by ChIP-PCR or ChIP-seq analysis. *Current Protocols in Microbiology.* 2016;40:1E 10 1–21. Epub 2016/02/09.

93. Baron L, Atar S, Zur H, Roopin M, Goz E, Tuller T. Computational based design and tracking of synthetic variants of Porcine circovirus reveal relations between silent genomic information and viral fitness. To appear in Scientific Reports. 2021.

94. National Institute of Health. Clinical Trials. 2020 [cited 2020 08/23/2020]; Available from: https://clinicaltrials.gov/ct2/results?cond=cancer&term=o ncolytic+virus&cntry=&state=&city=&dist=&Search=Search.

95. Eissa IR, Naoe Y, Bustos-Villalobos I, Ichinose T, Tanaka M, Zhiwen W, et al. Genomic signature of the natural oncolytic herpes simplex virus HF10 and its therapeutic role in preclinical and clinical trials. *Frontiers in Oncology.* 2017;7:149. Epub 2017/08/05.

96. Kasuya H, Kodera Y, Nakao A, Yamamura K, Gewen T, Zhiwen W, et al. Phase I dose-escalation clinical trial of HF10 oncolytic herpes virus in 17 Japanese patients with advanced cancer. *Hepato-Gastroenterology.* 2014; 61(131):599–605. Epub 2014/05/01.

97. Msaouel P, Dispenzieri A, Galanis E. Clinical testing of engineered oncolytic measles virus strains in the treatment of cancer: An overview. *Current Opinion in Molecular Therapeutics.* 2009;11(1):43–53. Epub 2009/01/27.

98. Galanis E, Atherton PJ, Maurer MJ, Knutson KL, Dowdy SC, Cliby WA, et al. Oncolytic measles virus expressing the sodium iodide symporter to treat drug-resistant ovarian cancer. *Cancer Research.* 2015;75(1):22–30. Epub 2014/11/16.

CHAPTER 13

Sharing Knowledge in Virology

Edouard De Castro, Chantal Hulo,
Patrick Masson, and Philippe Le Mercier
University of Geneva Medical School

CONTENTS

Virology deals with simple but much varied genomic entities. The apparent simplicity of their small genome is contradicted by the complexity of viral infections: a small genetic entity takes over the direction of a whole cell. It can even affect or kill an entire organism. The biology of some viruses like HIV is very well documented; however, for many exotic viruses, access to knowledge can be complicated or only accessible to experts. Fragmentation of knowledge is another peculiarity of virology and is due to the great diversity of virus entities: experts working on the influenza virus have very little interaction with those specialists in herpes viruses. Knowledge supports are traditionally encyclopedic books, publications, and websites often centered on one or a few viruses. The accessible data are heterogeneous in terms of both content and illustrations and their accessibility is sometimes limited: (i) books are expensive and slowly updated (fields virology, principles of virology, etc.) and (ii) peer-reviewed publications constitute a mine of information but remain highly heterogeneous, devoid of updates, and sometimes with paid access. On the other hand, websites and databases offer the most promising space for sharing knowledge in a sustainable and accessible way.

13.1 THE VIRUS EXCEPTION

Viruses have long been at the forefront of biology research: Phages study has proven that DNA is the carrier of genetic information [1], the discovery of reverse-transcriptase opened the study of RNA world [2], and phages Lambda or M13 were the foundation of 1980s genetic studies. This kind of virus leadership did not extend to viral bioinformatics: most resources or tools have been designed originally for cellular organisms and virus-specific tools have been created later. This results in data structure that is not always adapted to virus biology. Many viruses have a RNA genome; still they are all represented under DNA code in all sequence databases and not always clearly indicated as being RNA. One has to deduce it from taxonomic data. Moreover, viruses are parasitic organisms; they depend on the cell they infect. In other words, the viral entity is actually defined by the virus plus its host cell. This dual taxonomy is poorly handled by classical databases. INSDC database provides a TaxID for all species that are structured and searchable. Virus host is only defined by "vertebrate, plant, etc.," more specific host can be displayed under source /host with a scientific name, but this is not mandatory. Host data can be lost and cannot be fully exploited. However, this data is essential to make sense of virus sequences. At the level of protein products, the main issues come from viral polyproteins. These polypeptides are a means to express various proteins through a long chain that is subsequently cleaved. A viral polyprotein is a concatenation of several products, which would be encoded by several genes in other organisms. Because only viruses use polyprotein strategy to this extend, protein databases do not allow studying individual polyprotein products. Are you interested in retroviral reverse transcriptase? Databases or blast will offer you the whole POL or GAG-POL polyprotein to work with!

All these issues call for the creation of specialized databases and tools to exploit virus datasets. Over the years, many resources have appeared to help exploiting and storing virus data: some generalist like Virus pathogen resource (ViPR) [3], NCBI virus [4], GLUE resources [5], ViralZone [6], and Virushostnet [7] (see more in Ref. [8]); other more specialized like Description of plant viruses [9], Hepseq [10], and ACLAME [11]. These resources allow studying virus information with appropriate tools and data structure. These have proven invaluable to speed up the study and sharing information. During SARS-CoV-2 pandemic, Nextstrain has allowed monitoring the genetic evolution of the virus pretty much live [12], and many resources and databases have provided early bioinformatics tools designed explicitly for SARS-CoV-2 and coronaviruses research [13,14].

13.2 VIRALZONE AT THE SERVICE OF KNOWLEDGE SHARING

Digital technologies offer the opportunity to classify, maintain, and share knowledge in a very efficient way. ViralZone was created in this spirit: provide quality, consistent data that is freely accessible and regularly updated [6]. A general portrait of each family and viral genus have been established in ViralZone after extensive bibliographical research and with the help of experts. ViralZone site offers illustrations of virions for all virus genera, represented in a homogeneous and rigorous way. As far as possible, the virions are drawn from photos of transmission electron microscopy or cryo-EM (Figure 13.1) in order to preserve the real proportions of the

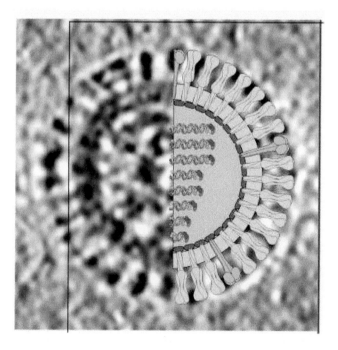

FIGURE 13.1   Representation of a viral particle of the influenza virus whose surface proteins and membrane are proportioned according to a cryo-EM photo [15].

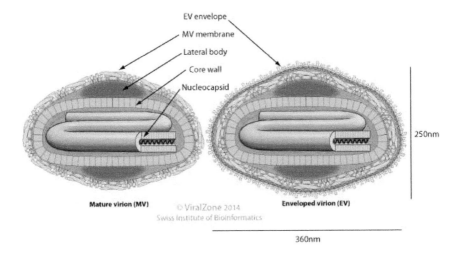

FIGURE 13.2 Vaccinia virus virions representation needed the advice of Bernard Moss and Rich Condit for integrating electron microscopy and biochemical data.

protein complexes, membranes, and capsid(s). Only the nucleic acids are not shown to scale since they would be too thin to be visible.

For some viruses, it is difficult to draw a virion conforming to all the details revealed in publications. For those, expert opinions have been needed to obtain satisfactory results. For example, the representation of the poxvirus virion (Figure 13.2) benefited from the advice of Rich Condit and Bernard Moss to represent accurately the viral particle and its envelope variants [16].

The genetic structure of viral genomes is also presented in ViralZone, with genes and segments named according to observed nomenclature in publications (Figure 13.3).

Official genomic nomenclatures exist for most cellular organisms (for humans it is defined by the Gene HUGO -HGNC Nomenclature Committee). This is not the case in virology: each community defines its own standards, or not. This can result in confusing names. Take, for example, the genomic nomenclature of coronaviruses. Cleavages products of the polyproteins (Open Reading Frames 1a and 1ab) were named non-structural-proteins (nsp), numbered 1–16. The problem is that the subsequent ORFs 3–9 were also called non-structural-proteins (nsp) and numbered 3–9. Nsp6, therefore, corresponds to two different proteins product according to this convention. This created a lot of confusion in the databases at the onset of COVID-19. The problem was solved with the help of

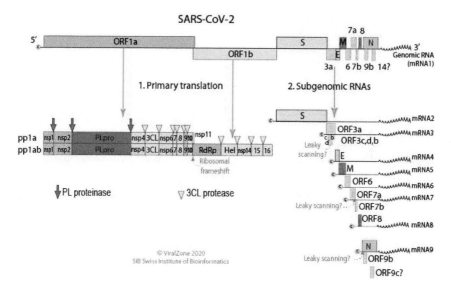

FIGURE 13.3 SARS-CoV-2 genome, transcripts, and gene products representation in ViralZone. Special features like polyprotein cleavages, ribosomal frameshift, and leaky scanning are represented.

experts who suggested naming protein products outside of the polyprotein "ORF" instead of "nsp" (Figure 13.3): The second nsp6 is now named ORF6. Thus, digitizing knowledge raised unexpected problems but improved data consistency at the end of the day.

The common education to all biologists is that of cellular biochemistry. However, viruses display many unique molecular activities that are not described in common molecular biology. This is why ViralZone also offers pages detailing the main viral processes. For example, the "ribosomal frameshift" page provides a description of the phenomenon which is part of "Viral replication/transcription/translation" section (Figure 13.4). There is no consensus to name these molecular mechanisms, and several naming conventions coexist: ribosomal frameshift=ribosomal frame shifting=translation frame shift. Yet, associating a precise vocabulary with data is an essential step in digitizing and exploiting knowledge because it allows fixing concepts that are sometimes named too loosely. A controlled vocabulary was therefore created to name all these virus-specific phenomena, which served as a model for establishing new Gene Ontology terms (GO terms) [17]. For example, reverse transcription of hepatitis B and HIV consist both in converting vRNA into vDNA, but the biochemical procedures involved are different for those viruses. We have created

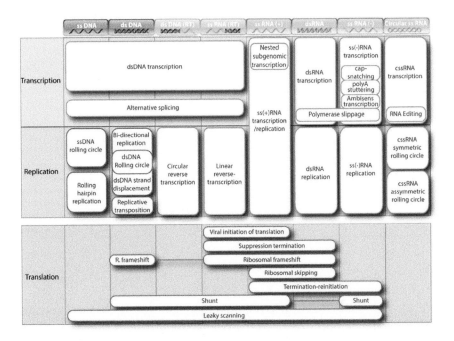

FIGURE 13.4 Transcription, replication, and translation process of the viral world. The processes shared with cellular organisms are circled in red. In the line "Replication," we find the reverse circular and linear transcription. Note the addition of circular ssRNAs to Baltimore classification, created to describe better Deltaviruses and viroids processes.

two terms to describe this difference: "Circular reverse transcription" for Hepadnaviridae (HBV), and "linear reverse transcription" for Retroviridae (HIV). Such viral controlled vocabulary has been developed in an exhaustive manner, and it facilitates understanding virology with hindsight. Thus, the diversity of viral transcription and translation processes appears clearly in view of Figure 13.4.

Overall, a global effort has been done in ViralZone to gather, share, and perpetuate general knowledge in virus molecular biology. The virus fact sheets are linked to many outstanding virus resources, ranging from sequence databases to phylogenies or other specific tools.

13.3 THE GROWING LANDSCAPE OF VIRUS DATABASES

Over the years, many specialized databases and tools have been created to describe virus biology. The main challenge for these resources is obtaining financial support for maintaining and updating their data.

TABLE 13.1   Non-Exhaustive List of Major Virus Resources

Database (Alphabetic Order)	Link	Focus
Aclame	http://aclame.ulb.ac.be/	Mobile genetic elements
GLUE	https://bioinformatics.cvr.ac.uk/projects/glue/	Data tools, information
ICTV	https://talk.ictvonline.org/	Virus taxonomy
INSDC	http://www.insdc.org/	Sequence raw data, taxonomy
NCBI virus	https://www.ncbi.nlm.nih.gov/labs/virus/vssi/#/	Curated references and raw sequences
Nextstrain	https://nextstrain.org/	Pathogenic virus phylogeny
UniProt	http://www.uniprot.org/	Curated proteins
VBRC	https://4virology.net/	Large nucleo-cytoplasmic viruses
VIPERdb	http://viperdb.scripps.edu/	3D capsid representation
ViPr	https://www.viprbrc.org/brc/home.spg?decorator=vipr	Curated sequences and tools
ViralZone	https://viralzone.expasy.org/	General knowledge
Virus-Host DB	https://www.genome.jp/virushostdb/	Host organisms

Unfortunately, public grants tend to fund more easily the acquisition of data than the storage and maintenance of it. Thus, "DPV: description of plant viruses" (https://www.dpvweb.net/) was an excellent site developed by Mike Adams, and the site is no longer updated since 2013. Some databases appear and disappear each year. Fortunately, there are major virus resources sustained in the long term (Table 13.1).

There are also virus databases dedicated to specific viruses, for example Influenzavirus Research Database (IRD https://www.fludb.org/brc/home.spg?decorator=influenza) [18], Hepatitis B virus database (HBVdb https://hbvdb.lyon.inserm.fr/HBVdb/) [19], Hepatitis C virus database (euHCVdb https://euhcvdb.lyon.inserm.fr/euHCVdb/) [20], or Global Initiative on Sharing Avian Influenza Data (GISAID https://www.gisaid.org/) [21]. These provide data structure and tools adapted to specific viruses and allow in deep study of its available data.

Does this database landscape cover all the needs of the scientific community? A huge effort has been done over the last 10 years, but there is room for improvement. Resources have been developed for the most prevalent pathogenic viruses. Nonetheless, it would be interesting to offer the same level of quality to all kinds of viruses, because new virus can emerge from any part of the virus taxonomy. Ideally, the same kind of web resource would be able to gather and exploit all viruses' data. ViPR is a good example of the ideal database [3], as it offers the same data structure for many viruses, with specific tools and links to molecular biology and clinical data. The database covers about 20 virus families, but at least 39 families would be necessary to cover all vertebrates' viruses. The coverage can certainly be improved, possibly by adding resources for missing viruses in a compatible format.

Moreover, addition of clinical data to the sequences database could be improved. Each pathogenic virus is linked to prevalence, incidence, incubation time, disease, fatality rate, etc. Each virus is replicating in certain organs and specific cell types. These data are not yet available and/or linked to viral databases, and giving access to it would be a major improvement.

13.4 THE PREDICTIVE POWER OF KNOWLEDGE

One of the most important aspects of knowledge is the power of prediction that is often formulated with Bayes' Theorem. Actually, many knowledge in virology comes from prediction, relying on similarity or models. Within a virus family, few strains are studied in laboratory and knowledge about other viruses is inferred by similarity. Moreover, understanding viral molecular biology allows deducing concepts within different families. For example, it is known that HIV-1 unspliced genomic RNA is exported from the nucleus thanks to viral rev protein [22]. Indeed cellular unspliced mRNAs are constrained inside the nucleus to avoid creating confusion in the cytoplasm. HBV virus is also a reverse transcribing virus that exports its unspliced genomic RNA, but no specific mechanism has been discovered to do so. Comparison with HIV allows predicting a missing link that deserves investigation and is currently still under active research [23]. Deep knowledge of a virus allows identifying missing links in the other. "Real knowledge is to know the extent of one's ignorance" (Confucius 500 B.C.).

How could expert knowledge be translated into resources amenable to be exploited by computation? Handling knowledge and making accurate predictions is a challenge of tomorrow's bioinformatics. In the head of experts, concepts do not have stable names or ID. Relations between concepts are complex and often unconscious. The challenge is to organize this into data structure that can be handled by computers. In bioinformatics, representation of knowledge relies on controlled vocabulary or ontologies [24], which is a concrete form of a conceptualization of a community's knowledge of a domain. Translating human knowledge into ontology is long and complex and requires a community effort. The main resource for biological ontologies is the Open Biological and Biomedical Ontology (OBO) Foundry [25], a collection of ontologies designed for biology and including gene ontology [26].

Modeling is at the crossroad between knowledge and predictions. A lot of research produces models for virus epidemiology, structure prediction, or antiviral discovery. Although modeling is complex to apply in biology, and the results are often rebuttable [27]. The predictive powers of bioinformatics models rely on the quality and completeness of data used to build predictors, which are not always fully available. For example, early predictions of protein secondary structure based on amino acidic sequence were rather unreliable, because too few structures were known and the technology was not mature enough [28]. Later on, SARS-COV-2 spike has been modeled accurately a few days after the release of its sequence; allowing the simulation of antiviral drugs docking [29]. More recently, 3D modeling has achieved solving protein structure accurately [30]. We are now close to predict very accurately 3D structures, but the journey has been long since the creation of PDB in 1971.

13.5 CONCLUSION

Digitalizing knowledge is essential to exploit the mighty power of prediction offered by bioinformatics, which can in turn convert data to knowledge. In the past 15 years, many efforts have been achieved toward this goal. The tremendous acceleration of bioinformatics research witnessed with the COVID-19 pandemic suggests we are on the right track. Nonetheless, there is still work to be done to achieve a complete adequacy. The full digitization of knowledge will offer unparalleled potential for researchers in the future. Anyone will be able to access and connect knowledge that previously required the acquisition of long expertise, and bioinformatics will offer unprecedented power to improve human knowledge.

REFERENCES

1. Hershey AD, Chase M. Independent functions of viral protein and nucleic acid in growth of bacteriophage. *J Gen Physiol*. 1952;36: 39–56. doi:10.1085/jgp.36.1.39.
2. Coffin JM, Fan H. The discovery of reverse transcriptase. *Annu Rev Virol*. 2016;3: 29–51. doi:10.1146/annurev-virology-110615-035556.
3. Pickett BE, Sadat EL, Zhang Y, Noronha JM, Squires RB, Hunt V, et al. ViPR: An open bioinformatics database and analysis resource for virology research. *Nucleic Acids Res*. 2012;40: D593–D598. doi:10.1093/nar/gkr859.

4. Hatcher EL, Zhdanov SA, Bao Y, Blinkova O, Nawrocki EP, Ostapchuck Y, et al. Virus variation resource: Improved response to emergent viral outbreaks. *Nucleic Acids Res.* 2017;45: D482–D490. doi:10.1093/nar/gkw1065.

5. Singer JB, Thomson EC, McLauchlan J, Hughes J, Gifford RJ. GLUE: A flexible software system for virus sequence data. *BMC Bioinf.* 2018;19: 532. doi:10.1186/s12859-018-2459-9.

6. Hulo C, de Castro E, Masson P, Bougueleret L, Bairoch A, Xenarios I, et al. ViralZone: A knowledge resource to understand virus diversity. *Nucleic Acids Res.* 2011;39: D576–D582. doi:10.1093/nar/gkq901.

7. Guirimand T, Delmotte S, Navratil V. VirHostNet 2.0: Surfing on the web of virus/host molecular interactions data. *Nucleic Acids Res.* 2015;43: D583–D587. doi:10.1093/nar/gku1121.

8. Tools. EVBC [Internet]. 24 Apr 2017 [cited 23 Sep 2020]. Available: http://evbc.uni-jena.de/tools/.

9. Adams MJ, Antoniw JF. DPVweb: A comprehensive database of plant and fungal virus genes and genomes. *Nucleic Acids Res.* 2006;34: D382–D385. doi:10.1093/nar/gkj023.

10. Gnaneshan S, Ijaz S, Moran J, Ramsay M, Green J. HepSEQ: International public health repository for hepatitis B. *Nucleic Acids Res.* 2007;35: D367–D370. doi:10.1093/nar/gkl874.

11. Leplae R, Hebrant A, Wodak SJ, Toussaint A. ACLAME: A classification of mobile genetic elements. *Nucleic Acids Res.* 2004;32: D45–D49. doi:10.1093/nar/gkh084.

12. Hadfield J, Megill C, Bell SM, Huddleston J, Potter B, Callender C, et al. Nextstrain: Real-time tracking of pathogen evolution. *Bioinformatics.* 2018;34: 4121–4123. doi:10.1093/bioinformatics/bty407.

13. Coronavirus Tools. EVBC [Internet]. 15 Apr 2020 [cited 23 Sep 2020]. Available: http://evbc.uni-jena.de/tools/coronavirus-tools/.

14. Hufsky F, Lamkiewicz K, Almeida A, Aouacheria A, Arighi C, Bateman A, et al. Computational strategies to combat COVID-19: Useful tools to accelerate SARS-CoV-2 and coronavirus research. 2020 [cited 23 Sep 2020]. doi:10.20944/preprints202005.0376.v1.

15. Harris A, Cardone G, Winkler DC, Heymann JB, Brecher M, White JM, et al. Influenza virus pleiomorphy characterized by cryoelectron tomography. *Proc Natl Acad Sci U S A.* 2006;103: 19123–19127. doi:10.1073/pnas.0607614103.

16. Condit RC, Moussatche N, Traktman P. In a nutshell: Structure and assembly of the vaccinia virion. *Adv Virus Res.* 2006;66: 31–124. doi:10.1016/S0065-3527(06)66002-8.

17. Hulo C, Masson P, de Castro E, Auchincloss AH, Foulger R, Poux S, et al. The ins and outs of eukaryotic viruses: Knowledge base and ontology of a viral infection. *PLoS One.* 2017;12: e0171746. doi:10.1371/journal.pone.0171746.

18. Zhang Y, Aevermann BD, Anderson TK, Burke DF, Dauphin G, Gu Z, et al. Influenza research database: An integrated bioinformatics resource for influenza virus research. *Nucleic Acids Res.* 2017;45: D466–D474. doi:10.1093/nar/gkw857.

19. Hayer J, Jadeau F, Deléage G, Kay A, Zoulim F, Combet C. HBVdb: A knowledge database for hepatitis B virus. *Nucleic Acids Res.* 2013;41: D566–D570. doi:10.1093/nar/gks1022.

20. Combet C, Garnier N, Charavay C, Grando D, Crisan D, Lopez J, et al. euHCVdb: The European hepatitis C virus database. *Nucleic Acids Res.* 2007;35: D363–D366. doi:10.1093/nar/gkl970.

21. Shu Y, McCauley J. GISAID: Global initiative on sharing all influenza data - from vision to reality. *Euro Surveill.* 2017;22. doi:10.2807/1560-7917. ES.2017.22.13.30494.

22. Felber BK, Hadzopoulou-Cladaras M, Cladaras C, Copeland T, Pavlakis GN. Rev protein of human immunodeficiency virus type 1 affects the stability and transport of the viral mRNA. *Proc Natl Acad Sci U S A.* 1989;86: 1495–1499. doi:10.1073/pnas.86.5.1495.

23. Makokha GN, Abe-Chayama H, Chowdhury S, Hayes CN, Tsuge M, Yoshima T, et al. Regulation of the hepatitis B virus replication and gene expression by the multi-functional protein TARDBP. *Sci Rep.* 2019;9: 8462. doi:10.1038/s41598-019-44934-5.

24. Stevens R, Goble CA, Bechhofer S. Ontology-based knowledge representation for bioinformatics. *Brief Bioinf.* 2000;1: 398–414.

25. Smith B, Ashburner M, Rosse C, Bard J, Bug W, Ceusters W, et al. The OBO foundry: Coordinated evolution of ontologies to support biomedical data integration. *Nat Biotechnol.* 2007;25: 1251–1255. doi:10.1038/nbt1346.

26. The Gene Ontology Consortium. The gene ontology resource: 20 years and still GOing strong. *Nucleic Acids Res.* 2019;47: D330–D338. doi:10.1093/ nar/gky1055.

27. Gelman A. Objections to Bayesian statistics. *Bayesian Anal.* 2008;3: 445–449. doi:10.1214/08-BA318.

28. Chou PY, Fasman GD. Prediction of protein conformation. *Biochemistry.* 1974;13: 222–245. doi:10.1021/bi00699a002.

29. Mothay D, Ramesh KV. Binding site analysis of potential protease inhibitors of COVID-19 using AutoDock. *Virusdisease.* 2020; 1–6. doi:10.1007/ s13337-020-00585-z.

30. Service RF. "The game has changed." AI triumphs at protein folding. *Science.* 2020;370: 1144–1145. doi:10.1126/science.370.6521.1144.

Index